explorations
in geometry

explorations
in geometry

bruce shawyer
memorial university of newfoundland, canada

World Scientific

NEW JERSEY · LONDON · SINGAPORE · BEIJING · SHANGHAI · HONG KONG · TAIPEI · CHENNAI

Published by

World Scientific Publishing Co. Pte. Ltd.

5 Toh Tuck Link, Singapore 596224

USA office: 27 Warren Street, Suite 401-402, Hackensack, NJ 07601

UK office: 57 Shelton Street, Covent Garden, London WC2H 9HE

British Library Cataloguing-in-Publication Data
A catalogue record for this book is available from the British Library.

ISBN-13 978-981-4295-85-7
ISBN-10 981-4295-85-X

Printed in Singapore.

I would like to mention several of the many people to whom I owe a debt of thanks.

First, my wife, Jo Shawyer, who has steadfastly been my best friend for many years.

Second, my mentor in mathematics, Dr. David Borwein of, first, the University of St. Andrews, and subsequently, of the University of Western Ontario. His guidance has been so valuable throughout my career.

Third, my Kirkcaldy High School Mathematics teacher, Mr. J.S. White, who instilled in me a love of geometry.

And last, but not least, my many colleagues and students who have made my career so enjoyable.

Many thanks are due to the students of the Fall 1993 class, who first used an early version of this book and who helped to eliminate many typographical errors. In particular I would like to thank Gerry McGrath and Trevor Young who handed in lists of such errors. Thanks are also due to students in subsequent classes for finding other typographical errors.

I would particulary like to thank my colleagues, Dr. Michael Parmenter and Mr. Afework Solomon, who have also taught the course, and have contributed considerably to the improvement of several sections; and also Dr. Danny Dyer, who used an earlier version of this book when at the University of Regina. Mr. Solomon was also of great assistance in the proofreading of the draft of this book.

Finally, I would like to express my sincere gratitude to my editor at World Scientific, Ms. Zhang Ji, for her excellent help in bringing this work into shape for publication.

Preface

It is said that Plato had inscribed above the entrance to his academy:

Let no-one unversed in Geometry enter here

I have always believed that a good grounding in Geometry gives one a good start for mathematical explorations. Solving geometrical problems involves experimentation. One has first to draw a diagram representing the given information, and then to add to this diagram in order to find a proof of the required result. This can lead to very messy diagrams, which one should then redraw with only the necessary constructions. I am constantly reminded of what I was taught in High School — **GRCP**:

(1) Given: (2) Required:
(3) Construction: (4) Proof.

The origins of this book lie in a course on Euclidean Geometry given at the third year level at Memorial University of Newfoundland, in St. John's.

Although the students in the course had studied Trigonometry and basic Euclidean Geometry in high school, I found that they had forgotten most of it, and so had to review it. They had no adequate source materials available to them, hence, the genesis of this book. This book contains over 200 problems, of which, most have solutions given.

Added to the work are some modern results. The sections on Remarkable Concurrencies and Remarkable Bisections contain some of my own work.

This book should prove useful to students at high school and undergraduate levels who have an interest in the various mathematics competitions. These competitions act as a stimulus to many students, who enjoy problem solving. I have been involved in helping in such events for many years, and still find it such an enjoyable occasion when students are having enjoyment solving mathematical problems.

Bruce Shawyer, Professor Emeritus
Department of Mathematics and Statistics
Memorial University of Newfoundland, St. John's,
Newfoundland and Labrador, Canada.

Contents

Chapter 1

Basic Euclidean Geometry

This chapter is not intended to be a complete survey of basic Euclidean Geometry, but rather a review for those who have previously taken a geometry course. For a definitive account, see Euclid's *Elements*.

1.1 Triangles

A triangle is a (plane) figure bounded by three line segments. The most important result about triangles is that the sum of the angles of a triangle has measure equal to two right angles (or 180°). This can be deduced from Fig. 1.1, where the line DAE is parallel to the line segment BC.

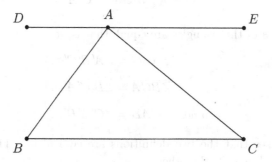

Fig. 1.1

If one of the angles of a triangle is a right angle, we call the the triangle a *right triangle*. For such triangles, we have *Pythagoras's Theorem* which states that the square of the hypotenuse (the side opposite the right angle) is equal to the sum of the squares of the other two sides.

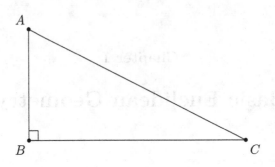

Fig. 1.2

In Fig. 1.2, we have that $AB^2 + BC^2 = CA^2$.

1.2 Similar Triangles

We start with two triangles ABC and $A'B'C'$. The definition of two triangles being similar can be given in one of two ways:
We say that the triangles ABC and $A'B'C'$ are *similar* if either

(1) the *sides* of the triangle are in proportion; that is, if

$$\frac{AB}{A'B'} = \frac{BC}{B'C'} = \frac{CA}{C'A'} \ ;$$

or

(2) the *angles* of the triangles are equal; that is, if

$$\angle ABC = \angle A'B'C' \ ,$$

$$\angle BCA = \angle B'C'A' \ ,$$

$$\text{and} \quad \angle CAB = \angle C'A'B' \ .$$

It is easy to see that the two definitions are equivalent. Thus, showing either relationship gives the other.
We ask the question: what is sufficient to show that two triangles are similar? Do we have to show that all three angles are equal; do we have to show that all three ratios are equal?
In the case of *angles*, it is sufficient to show that *two* of the angles are equal, since this will automatically give that the third angles are equal. (Why?)

However, it is not sufficient to show that two of the ratios of the sides are equal. To see this, simply consider two isosceles triangles, ABC and $A'B'C'$ with $AB = BC$ and $A'B' = B'C'$.

Then, $\frac{AB}{A'B'} = \frac{BC}{B'C'}$, but $\triangle ABC$ and $\triangle A'B'C'$ will only be similar if $\angle ABC = \angle A'B'C'$.

Problem 1.1. *Let ABC be a triangle. Points B' and C' are chosen on AB and AC, respectively, such that $\frac{AB'}{B'B} = \frac{AC'}{C'C}$. Prove that triangles ABC and AB'C' are similar.*

1.3 Congruent Triangles

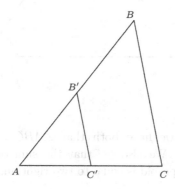

Similar triangles can be of different sizes. (We see that in the diagram above.) If the ratio of the sides has value 1, then we say that the triangles are *congruent*. This means that the two triangles are *identical* in every way, although their orientation and position may differ.

There are several conditions that are sufficient for showing that two triangles are congruent. They are

1. Three sides equal **SSS**

2. Two sides and the included angle **SAS**

3. Two angles and the corresponding side **ASA**

In the third case, we may as well assume that the side is common to both angles; hence, the notation **ASA**.

We do not insist on the same orientation for congruence.

1.4 Quadrilaterals

A quadrilateral is a plane figure bounded by four line segments.
In this section we look briefly at some particular quadrilaterals.

(1) **Trapezoid**

A trapezoid is a quadrilateral with one pair of (opposite) sides parallel.
a *symmetric* trapezoid is a trapezoid where the non-parallel sides are
equally inclined to the other sides. See Fig. 1.3.

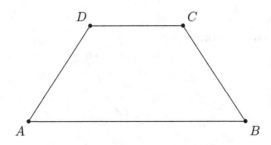

Fig. 1.3

In this example, we have both that $\angle ABC = \angle BAD$ and that
$\angle ADC = \angle BCD$. We also see that the sum of the opposite angle
of a symmetric trapezoid is equal to two right angles (π).

(2) **Parallelogram**

A parallelogram is a quadrilateral where the opposite sides are parallel.
This gives the result that opposite angle are equal. See Fig. 1.4.

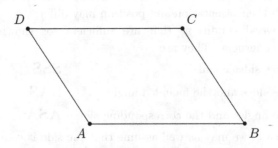

Fig. 1.4

(3) **Rhombus**

A rhombus is a parallelogram where all sides are equal.

(4) **Square**

A square is a particular quadrilateral where all sides are equal and all angles are equal to one right angle $(\pi/2)$.

Alternatively, a square is a rhombus where an angle is a right angle (which gives all angles as right angles).

1.5 Polygons

A polygon is a plane figure bounded by line segments. Some special names are:

Sides	Name
3	Triangle
4	Quadrilateral
5	Pentagon
6	Hexagon
n	Polygon

A basic result concerns the sum of the interior angles of a polygon. The value, $(2n-4)$ right angles, or $(n-2)\pi$, can be seen easily by triangulation: see Fig. 1.5.

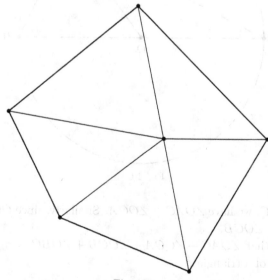

Fig. 1.5

Here we have n triangles. The total measure of the angles is then $n \times \pi$. From this, we must subtract the sum of the angles at the interior point, that is, 2π.

We shall be interested in a later chapter in *regular* polygons, and which can be constructed using straight edge and compasses.

1.6   Circles and Angles

In this section, we will recall a number of important results concerning angles, which arise naturally in the study of circles. The first concerns the size of an angle in a semicircle.

Theorem 1.1. *If AB is the diameter of a circle and C is any point on the circle distinct from A and B, then $\angle ACB = \pi/2$ (in radian measure).*

Proof.
Let O be the centre of the circle. to help, join OC, AC and CB. See Fig. 1.6.

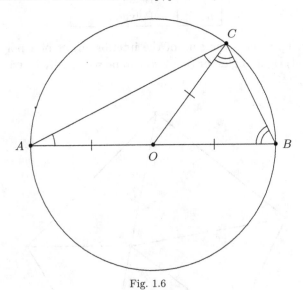

Fig. 1.6

Since $OA = OC$, we have $\angle OAC = \angle OCA$. Similarly, since $OC = OB$, we have $\angle OBC = \angle OCB$.

But we know that $\angle OAC + \angle OCA + \angle OCB + \angle OBC = \pi$ (sum of the interior angles of a triangle).

Hence, $2\angle OCA + 2\angle OCB = \pi$, giving $\angle OCA + \angle OCB = \pi/2$. Therefore, $\angle ACB = \pi/2$ as desired.

Next, we consider the relationship between the angle subtended by an arc to a point on the circumference with the angle subtended by the same arc at the centre.

Theorem 1.2. *Let AB be a chord of a circle (with centre O) which is not a diameter, and let C be any point on the circle distinct from A and B.*

(1) If C is on the same side of AC as O, then $\angle AOB = 2\angle ACB$.
(2) If C is on the opposite side of AB from O, then $\angle AOB = 2\pi - 2\angle ACB$.

Proof.
Let C be on the same side of AB as O. To help, join AO, OB, AC and CB. First, assume that AC does not intersect the radius OB, and that BC does not intersect the radius OA as illustrated in Fig. 1.7.

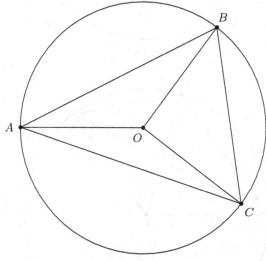

Fig. 1.7

Here, we have $\angle AOB = \pi - (\angle OAB + \angle OBA)$.
We also have $\angle OAC + \angle OCA + \angle OBC + \angle OCB = \pi - (\angle OAB + \angle OBA)$.
Hence, $\angle AOB = \angle OAC + \angle OCA + \angle OCB + \angle OBC$.
Since $OA = OC$, we have $\angle OAC = \angle OCA$, and since $OB = OC$, we have $\angle OBC = \angle OCB$.
Thus, $\angle OAB = 2(\angle OCA + \angle OCB) = 2\angle ACB$.

Next, assume that AC intersects the radius OB as illustrated below (the case where BC intersects OA is proved similarly, and is left as an exercise): see Fig. 1.8.

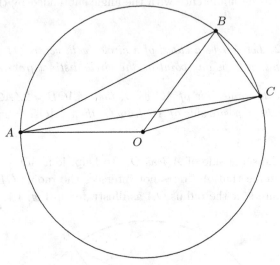

Fig. 1.8

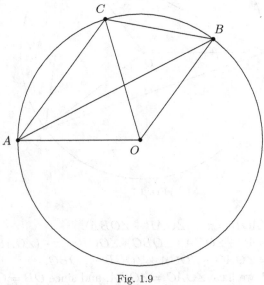

Fig. 1.9

Again, we have $\angle AOB = \pi - (\angle OAB + \angle OBA)$.

But this time, we have

$$(\angle OAB - \angle OAC) + (\angle OCB - \angle OCA) + (\angle OBA + \angle OBC) = \pi.$$

Thus, $\angle AOB = -\angle OAC + \angle OCB - \angle OCA + \angle OBC$.

As before, we have $\angle OAC = \angle OCA$ and $\angle OBC = \angle OCB$.

Hence, $\angle AOB = 2\angle OCB - 2\angle OCA = 2\angle ACB$.

Now, let C be on the opposite side of AB from the centre, and again, join AO, OB, AC, CB and OC. See Fig. 1.9.

Here, we have $\angle AOB = 2\pi - (\angle OAC + \angle OCA + \angle OCB + \angle OBC)$.

Since $OA = OC$, we have $\angle OAC = \angle OCA$, and since $OB = OC$, we have $\angle OBC = \angle OCB$.

Thus, $\angle AOB = 2\pi - 2\angle OCA - 2\angle OCB = 2\pi - 2\angle ACB$ as desired.

The next result, which follows immediately from theorem 1.2, will often be more important in applications.

Theorem 1.3. *If AB is a chord of a circle and C and D are two distinct points on the circle, distinct from A and B, both of which lie on the same side of AB, then $\angle ACB = \angle ADB$.*

Proof.

If AB is a diameter, then the result follows from theorem 1.1, since both $\angle ACB$ and $\angle ADB$ are right angles.

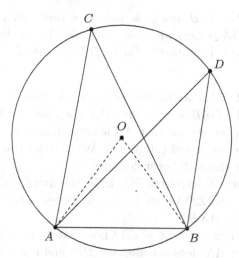

Fig. 1.10   The "Bow-Tie" Lemma

Assume, from now on, that AB is not a diameter. Let O be the centre of the circle.

If C and D are on the same side of AB as O, then theorem 1.2 (1) gives that $2\angle ACB = \angle AOB$ and $2\angle ADB = \angle AOB$, yielding that $\angle ACB = \angle ADB$. If C and D are on the opposite side of AB from O, then theorem 1.2 (2) enables us also to conclude that $\angle ACB = \angle ADB$.

If C and D are on opposite sides of AB, we have the following:

Theorem 1.4. *If AB is a chord of a circle and C and D are two points on the circle, distinct for A and B, lying on opposite sides of AB, then $\angle ACB + \angle ADB = \pi$.*

Proof.
As in the previous result, if AB is a diameter, then
$$\angle ACB + \angle ADB = \pi/2 + \pi/2 = \pi.$$
Assume for the remainder of this proof that AB is not a diameter, and let O be the centre of the circle. Without loss of generality, assume that C and O lie on the same side of AB. Then, theorem 1.2 tells us that $\angle AOB = 2\angle ACB$ and that $\angle AOB = 2\pi - 2\angle ADB$.

Hence, $2\angle ACB = 2\pi - 2\angle ADB$, yielding that $\angle ACB + \angle ADB = \pi$.

Finally, we will make an interesting observation involving the angles between chords and tangents.

Theorem 1.5. *Let A, B and C be any three points on a circle, and let D be such that DA is tangent to the circle (at A) and that D is on the opposite side of line AB from C. Then $\angle DAB = \angle ACB$.*

Proof.
First note that if AB is a diameter of the circle, then $\angle ACB = \pi/2$ by theorem 1.1, and $\angle DAB = \pi/2$ since DA is a tangent. This, the result holds. Henceforth, we assume that AB is not a diameter.

Draw the diameter at A and call it AX. We consider two cases — either AX intersects the chord BC or it does not.

First, assume that AX intersects BC. Draw the chord BX. See Fig. 1.10. Note that $\angle ABX$ and $\angle DAX$ are both right angles. Also, by theorem 1.3, we have $\angle ACB = \angle AXB$.

Hence, $\angle DAB = \pi/2 - \angle BAX = \angle AXB = \angle ACB$, as desired.

Next, assume that AX does not intersect BC, and again, draw the chord BX. See Fig. 1.11.

As above, $\angle ABX$ and $\angle DAX$ are right angles.

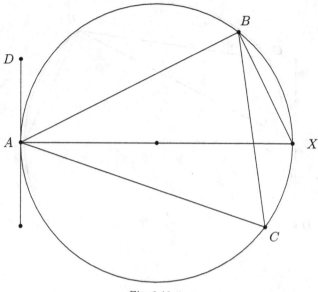

Fig. 1.11

In this case, $\angle ACB + \angle AXB = \pi$ by theorem 1.4. Hence,

$$\angle DAB = \pi/2 + \angle BAX = \pi/2 + (\pi/2 - \angle AXB)$$
$$= \pi - (\pi - \angle ACB) = \angle ACB.$$

1.7 Cyclic Quadrilaterals

A quadrilateral whose vertices all lie on a circle is called a **cyclic quadrilateral**.

Now, any triangle has the property that a unique circle can be drawn through its vertices (for the demonstration, theorem 3.6 on page 55). Thus, to say that a quadrilateral is cyclic is really quite restrictive — in fact, we are demanding that D lie on the unique circle passing through A, B and C. Similarly, we could start with the unique circle through any other three of the four named points. Therefore, we should expect that cyclic quadrilaterals have some very particular properties.

One such property is clear — from the geometry of a circle, we see that each angle of a cyclic quadrilateral $ABCD$ must be less than π. See Fig. 1.12.

To facilitate terminology, we call any quadrilateral with this property **convex** — in fact, there is a more general definition of convex which applies

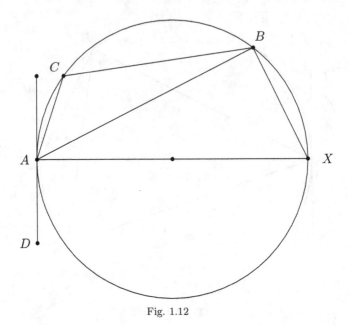

Fig. 1.12

to all polygons, but in the case of quadrilaterals, it is equivalent to ours.
Two other important properties of cyclic quadrilaterals follow immediately
from other result proved earlier. In fact, theorems 1.3 and 1.4 can be
re-stated as follows (we assume the vertices to be labelled in a clockwise
order):

(1) If $ABCD$ is a cyclic quadrilateral, then $\angle ACB = \angle ADB$.
(2) If $ABCD$ is a cyclic quadrilateral, then $\angle ABC + \angle ADC = \pi$.

Note that the first of these results also has three other possible conclusions,
depending on which of BC, CD or DA is chosen as the chord in theorem 1.3.
For example, $\angle DBA = \angle DCA$. As an exercise, the reader should list all
possibilities.
Similarly to the second result, we also have $\angle BCD + \angle BAD = \pi$. But this
could also be deduced from the fact that the sum of the interior angles of
a quadrilateral is 2π.
It is important for subsequent material to note that the converses of these
two results hold.

Theorem 1.6. *Suppose that $ABCD$ is a convex quadrilateral such that
$\angle ACB = \angle ADB$. Then quadrilateral $ABCD$ is a cyclic quadrilateral.*

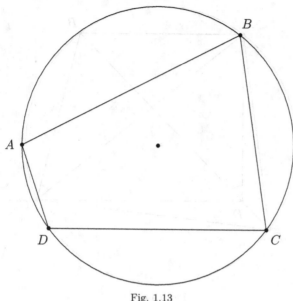

Fig. 1.13

Proof.

Draw the unique circle through A, B and C. We wish to prove that D lies on the circle. See Fig. 1.13.

Let E be the second point of intersection of the circle with BD (B being the first) and join AE. Since $ABCD$ is convex, the point E does exist and is on the same side of AC as is D. It follows that if $D \neq E$, then $\angle AEB \neq \angle ADB$. But, $ABCD$ is a cyclic quadrilateral, and thus, we have $\angle AEB = \angle ACB$. Since $\angle ACB = \angle ADB$ (as marked), this is a contradiction. Hence, $D = E$.

Theorem 1.7. *If $ABCD$ is a quadrilateral in which $\angle ABC + \angle ADC = \pi$, then $ABCD$ is cyclic.*

Proof.

Note that the given condition forces quadrilateral $ABCD$ to be convex. See Fig. 1.14.

As with theorem 1.6, draw the unique circle through A, B and C, and let E be the second point of intersection of the circle with BD. Join AE and AC. If $D \neq E$, then $\angle AEC \neq \angle ADC$. But $ABCE$ is cyclic, yielding that $\angle ABC + \angle AEC = \pi$. This implies that $\angle AEC = \angle ADC$, which is a contradiction.

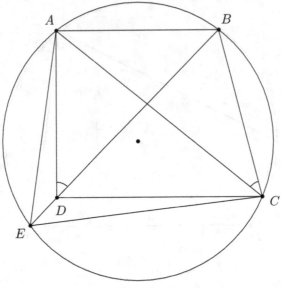

Fig. 1.14

1.8   Intersecting Chords

We consider a circle with two intersecting chords.

Suppose that AC and BD are two chords of a given circle which intersect at some point X inside the circle. See Fig. 1.15.

Here, we observe that theorem 1.3 implies that $\angle ABD = \angle ACD$, and thus, it follows that triangles ABX and DCX are similar. Hence, we have that $\frac{AX}{XD} = \frac{BX}{XC}$, or, equivalently,

$$AX \cdot XC = BX \cdot XD \ .$$

This result is known as the **Intersecting Chords Theorem**.

Of course, it can be immediately interpreted as a result about the diagonals of any cyclic quadrilateral.

Problem 1.2. *In equilateral triangle ABC of side length 2, suppose that M and N are the mid-points of AB and AC, respectively. The triangle is inscribed in a circle. The line segment MN is extended to meet the circle at P. Determine the length of the line segment NP.* (Solution on page 157.)

Problem 1.3. *Prove the converse of the Intersecting Chords Theorem. That is, prove that if $ABCD$ is a convex quadrilateral, if X is the point of*

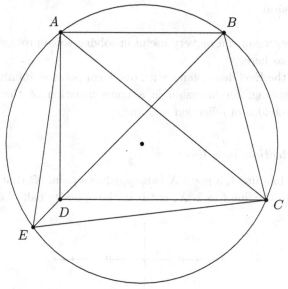

Fig. 1.15

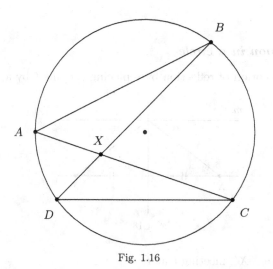

Fig. 1.16

intersection of the diagonals AC and BD, and if $AX \cdot XC = DX \cdot XB$, then $ABCD$ is a cyclic quadrilateral.

1.9   Inversion

The use of inversion can be very useful in solving some problems. We give
the basic ideas here.
We work in the Euclidean plane, with one ideal point at infinity ∞.
Roughly speaking, an inversion is a transformation of the plane that
generalizes the idea of reflection in a line.

1.9.1   *Reflection in a line*

In reflection in a line ℓ, a point X is mapped to a point X' that is the same
distance from the line ℓ as is X, but is in the opposite half plane to X.

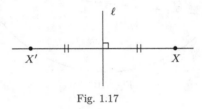

Fig. 1.17

1.9.2   *Inversion in a circle*

Generalize this notion of reflection by replacing the line ℓ by a circle Γ.

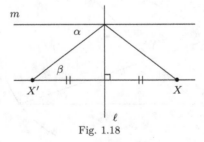

Fig. 1.18

Suppose that $m\|XX'$, meeting ℓ at P.
Reflecting $\angle PX'X$ gives $\angle PXX'$, which are equal.
Since $m\|XX'$, we have $\alpha = \beta$, so that $\angle PXX' = \alpha = \beta$.
Now, let your imagination expand so that ℓ is an infinitely large circle, with
m lying on the radius of this circle through the point P. We now view is
thus:

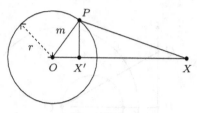

Fig. 1.19

We have $\triangle POX'$ similar to $\triangle XOP$, giving $\frac{OX'}{OP} = \frac{OP}{OX}$, so that $OX \cdot OX' = OP^2 = r^2$.

Now, suppose that O is a fixed point (called the CENTRE OF INVERSION) and that c is a fixed positive number (called the RADIUS OF INVERSION).

Definition 1.1. P and Q are INVERSES with respect to O with radius c if

$$OP.OQ = c^2 \ .$$

Definition 1.2. The circle centre O and radius c is called the CIRCLE OF INVERSION.

Theorem 1.8. *The circle of inversion is invariant under inversion.*

Theorem 1.9. *The inverse of a line through the centre of inversion is that line. (But it is not an invariant.)*

The proofs of these theorems are left to the reader.

Theorem 1.10. *The inverse of a line, not through the centre of inversion, is a circle passing through the centre of inversion.*

Proof.
Let O be the centre of inversion and PQ be the line, such that $OP \perp PQ$.
Let Γ be the circle of inversion.
Let P' and Q' be the inverses of P and Q, respectively.
Then $OP.OP' = OQ.OQ'$, giving

$$\frac{OP}{OQ} = \frac{OP'}{OQ'} \ .$$

Thus, triangles $\triangle OPQ$ and $\triangle OQ'P'$ are similar.
Since $\angle QPO = 90°$, we have that $\angle P'Q'O = 90°$, giving that O, P' and Q' lie on a circle of diameter OP'.
THE CONVERSE IS ALSO TRUE.

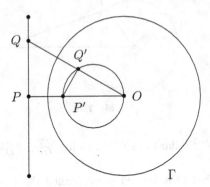

Fig. 1.20   Theorem 1.10

Theorem 1.11. *The inverse of a circle not passing through the centre of inversion in a circle (not passing through the centre of inversion).*

Proof.

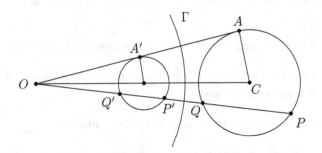

Fig. 1.21   Theorem 1.11

Let O be the centre of inversion and let the given circle have centre C. Then $OP.OP' = OQ.OQ' = c^2$.

Also, note that $OP.OQ = OA^2 = k^2$, where OA is the tangent from O to the given circle.

Thus,

$$c^4 = (OP.OP').(OQ.OQ') = (OP'.OQ').(OP.OQ) = (OP'.OQ').k^2,$$

whence,

$$OP'.OQ' = \frac{k^2}{c^4} = (OA')^2.$$

This means that the image is a circle!

Now, here is a result for you to try to obtain yourself!

Problem 1.4. *The measure of the angle between two intersecting circles is invariant under inversion*

Here is a useful figure.

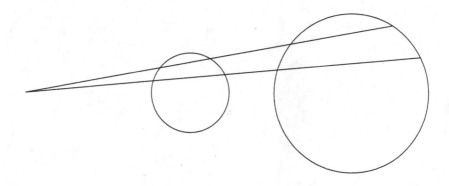

Problem 1.5. *Two points, A and B are distinct and not collinear with the centre O of the circle of inversion. The images of the two points are A′ and B′, respectively.*
Prove that triangles OAB and OB′A′ are similar. Note that their orientations are reversed.

Problem 1.6. *A straight line passing through the centre of the circle of inversion maps onto itself.*

Chapter 2

Trigonometry

2.1 Basic Definitions

We start with a right triangle, ABC, with $\angle ACB$ as the right angle. We shall denote the angle $\angle BAC$ by θ. The unit of measure for angles will be **radian** measure. See Fig. 2.1.

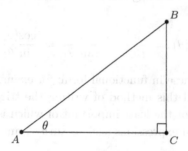

Fig. 2.1 Radian measure

The three main trigonometric functions are defined as follows:

Definition 2.1. The **sine** of the angle θ in the above triangle is defined by

$$\sin(\theta) = \frac{BC}{AB}$$

Definition 2.2. The **cosine** of the angle θ in the above triangle is defined by

$$\cos(\theta) = \frac{AC}{AB} \cdot$$

Definition 2.3. The **tangent** of the angle θ in the above triangle is defined by

$$\tan(\theta) = \frac{BC}{AC} = \frac{\sin(\theta)}{\cos(\theta)} \; .$$

There are also the three other trigonometric functions, defined as follows:

Definition 2.4. The **secant** of the angle θ in the above triangle is defined by

$$\sec(\theta) = \frac{AB}{AC} = \frac{1}{\cos(\theta)} \; .$$

Definition 2.5. The **cosecant** of the angle θ in the above triangle is defined by

$$\csc(\theta) = \frac{AB}{BC} = \frac{1}{\sin(\theta)} \; .$$

Definition 2.6. The **cotangent** of the angle θ in the above triangle is defined by

$$\cot(\theta) = \frac{AC}{BC} = \frac{1}{\tan(\theta)} = \frac{\cos(\theta)}{\sin(\theta)} \; .$$

Note that we write these in functional form; for example, $\sin(\theta)$ instead of $\sin \theta$. We recommend this method of writing the trigonometric functions for several reasons, not the least important of which is that it reduces the confusion arising from expressions such as $\sin \theta \cdot 2$ and $(\sin \theta)^{-1}$.

Problem 2.1. *Prove the following equalities:* (Solution on page 159.)

(1) $\sin^2(\theta) + \cos^2(\theta) = 1$;
(2) $\tan^2(\theta) - \sec^2(\theta) = -1$;
(3) $\cot^2(\theta) - \csc^2(\theta) = -1$.

2.2   Complementary Angles

Definition 2.7. Two angles are said to be **complementary** if their **sum** has measure one right angle (or $\pi/2$).
The **complement** of an angle of measure θ is the angle of measure $\pi/2 - \theta$.

In the given right triangle, it is obvious that the complement of the angle $\theta = \angle BAC$ is $\angle ABC$. See Fig. 2.2.

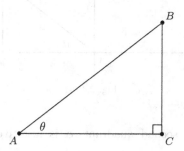

Fig. 2.2   Complement of an angle

It is now easy to see that

$$\sin(\pi/2 - \theta) = \cos(\theta) \; ; \qquad \cos(\pi/2 - \theta) = \sin(\theta) \; ;$$
$$\tan(\pi/2 - \theta) = \cot(\theta) \; .$$

2.3   Supplementary Angles

Definition 2.8. Two angles are said to be **supplementary** if their **sum** has measure two right angles (or π).
The **supplement** of an angle of measure θ is the angle of measure $\pi - \theta$. See Fig. 2.3.

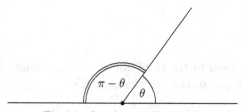

Fig. 2.3   Supplement of an angle

If θ is greater than $\pi/2$, we have angles that cannot be in right triangles. We will therefore consider quadrants. See Fig. 2.4.
If (as usual) we consider the polar distance (AB) as being positive, and associate the coordinates (α, β) with the point B, we see that the signs of α and β will vary with the quadrant in which B lies.

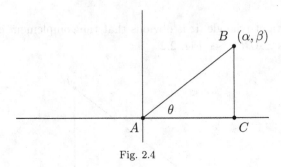

Fig. 2.4

For example, in the second quadrant, α is negative and β is positive. See Fig. 2.5.

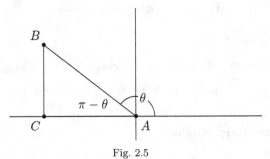

Fig. 2.5

From this, we interpret $\sin(\theta)$ as $\frac{|BC|}{|AB|}$, and therefore, is a positive quantity. A similar interpretation shows that $\cos(\theta)$ is $\frac{-|AC|}{|AB|}$; that is, a negative quantity.

2.4 Extensions

Extending these ideas to the third and fourth quadrants, shows that, for the three basic trigonometric functions, we have:

Quadrant	sin	cos	tan
First	+	+	+
Second	+	−	−
Third	−	−	+
Fourth	−	+	−

This is often remembered as "**A**ll, **S**ine, **T**angent, **C**osine", or as "**A**ll **S**tudents **T**ake **C**alculus".

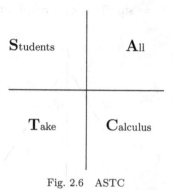

Students | **A**ll

Take | **C**alculus

Fig. 2.6   ASTC

Another version is: "**A**ll **S**cience **T**eachers are **C**razy".

Problem 2.2. *Give the following in the form:* $\pm \sin(\theta)$ *or* $\pm \cos(\theta)$:
(Solution on page 158.)

$\sin(\pi - \theta)$ $\sin(\pi + \theta)$ $\sin(2\pi - \theta)$ $\sin(2\pi + \theta)$
$\cos(\pi - \theta)$ $\cos(\pi + \theta)$ $\cos(2\pi - \theta)$ $\cos(2\pi + \theta)$
$\sin(\pi/2 - \theta)$ $\sin(\pi/2 + \theta)$ $\sin(3\pi/2 - \theta)$ $\sin(3\pi/2 + \theta)$
$\cos(\pi/2 - \theta)$ $\cos(\pi/2 + \theta)$ $\cos(3\pi/2 - \theta)$ $\cos(3\pi/2 + \theta)$

Problem 2.3. *Give the following in the form:* $\pm \tan(\theta)$ *or* $\pm \cot(\theta)$:
(Solution on page 158.)

$\tan(\pi - \theta)$ $\tan(\pi + \theta)$ $\tan(2\pi - \theta)$ $\tan(2\pi + \theta)$
$\cot(\pi - \theta)$ $\cot(\pi + \theta)$ $\cot(2\pi - \theta)$ $\cot(2\pi + \theta)$
$\tan(\pi/2 - \theta)$ $\tan(\pi/2 + \theta)$ $\tan(3\pi/2 - \theta)$ $\tan(3\pi/2 + \theta)$
$\cot(\pi/2 - \theta)$ $\cot(\pi/2 + \theta)$ $\cot(3\pi/2 - \theta)$ $\cot(3\pi/2 + \theta)$

It is also clear how to extend beyond the usual range of θ, since $\theta + 2k\pi$ represents that same angle as θ.

Problem 2.4. *Give the following in terms of* $\sin(\theta)$, $\cos(\theta)$ *or* $\tan(\theta)$:
(Solution on page 159.)

$\sin(-\theta)$ $\cos(-\theta)$ $\tan(-\theta)$ $\cot(-\theta)$
$\sin(2\pi + \theta)$ $\cos(2\pi + \theta)$ $\tan(2\pi + \theta)$ $\cot(2\pi + \theta)$
$\sin(3\pi + \theta)$ $\cos(4\pi + \theta)$ $\tan(5\pi + \theta)$ $\cot(6\pi + \theta)$
$\sin(3\pi - \theta)$ $\cos(3\pi/2 - \theta)$ $\tan(7\pi - \theta)$ $\cot(9\pi/2 - \theta)$

2.5   Addition Formulae

Here, we show that (provided that $\theta > 0$, $\phi > 0$ and $\theta + \phi < \frac{\pi}{2}$),

$$\sin(\theta + \phi) \;=\; \sin(\theta)\cos(\phi) + \cos(\theta)\sin(\phi) \;\;.$$

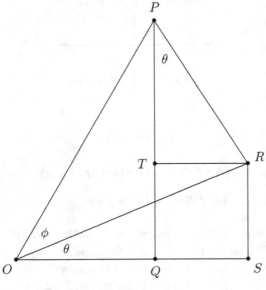

Fig. 2.7

In Fig. 2.7, we are given $\angle SOR = \theta$ and $\angle ROP = \phi$.
Let $OP = r$, $RS = QT = u$, $OR = w$, $TP = s$ and $PR = v$.
Note that $\angle OQP = \angle ORP = \pi/2$, so that quadrilateral $OQRP$ is cyclic.
See Chapter 1, page 11. Thus, $\angle QPR = \angle ROS = \theta$.
We have

$$\sin(\theta) = \frac{RS}{RO} = \frac{u}{w} \quad \cos(\theta) = \frac{PT}{PR} = \frac{s}{v}$$

$$\sin(\phi) = \frac{PR}{PO} = \frac{v}{r} \quad \cos(\phi) = \frac{OR}{OP} = \frac{w}{r}$$

so that

$$\sin(\theta)\cos(\phi) + \cos(\theta)\sin(\phi) = \frac{u}{w} \cdot \frac{w}{r} + \frac{s}{v} \cdot \frac{v}{r} = \frac{u+s}{r}$$

$$= \frac{PQ}{PO} = \sin(\theta + \phi) \;\;.$$

Problem 2.5. *Prove the following equalities:* (Solution on page 159.)

(1) $\sin(\theta - \phi) = \sin(\theta)\cos(\phi) - \cos(\theta)\sin(\phi)$;

(2) $\cos(\theta + \phi) = \cos(\theta)\cos(\phi) - \sin(\theta)\sin(\phi)$;

(3) $\cos(\theta - \phi) = \cos(\theta)\cos(\phi) + \sin(\theta)\sin(\phi)$;

(4) $\tan(\theta + \phi) = \frac{\tan(\theta) + \tan(\phi)}{1 - \tan(\theta)\tan(\phi)}$;

(5) $\tan(\theta - \phi) = \frac{\tan(\theta) - \tan(\phi)}{1 + \tan(\theta)\tan(\phi)}$.

2.6 Double Angle Formulae

From the above, we deduce the double angle formulae:

$$\sin(2\theta) = 2\sin(\theta)\cos(\theta) \; ;$$

$$\cos(2\theta) = \cos^2(\theta) - \sin^2(\theta)$$

$$= 2\cos^2(\theta) - 1 \; = \; 1 - 2\sin^2(\theta) \; ;$$

$$\tan(2\theta) = \frac{2\tan(\theta)}{1 - \tan^2(\theta)} \; .$$

Problem 2.6. *Find other proofs by searching in books.*

Making use of

$$\sin(\theta + \phi) = \sin(\theta)\cos(\phi) + \cos(\theta)\sin(\phi) \; ,$$

and $\quad \sin(\theta - \phi) = \sin(\theta)\cos(\phi) - \cos(\theta)\sin(\phi)$;

we get

$$\sin(\theta + \phi) + \sin(\theta - \phi) = 2\sin(\theta)\cos(\phi) \; .$$

and $\quad \sin(\theta + \phi) - \sin(\theta - \phi) = 2\cos(\theta)\sin(\phi)$;

from which we obtain

$$\sin(A) + \sin(B) = 2\sin\left(\frac{(A+B)}{2}\right)\cos\left(\frac{(A-B)}{2}\right) \; ,$$

and

$$\sin(A) - \sin(B) = 2\cos\left(\frac{(A+B)}{2}\right)\sin\left(\frac{(A-B)}{2}\right) \; .$$

Problem 2.7. *Find formulae for*

$$\cos(A) + \cos(B) \; , \quad \cos(A) - \cos(B) \; .$$

What can you do for $\tan(A) + \tan(B)$ *?* (Solution on page 161.)

2.7   Half Angle Formulae

Let $t = \tan\left(\frac{\theta}{2}\right)$.

Then $\qquad 1 + t^2 = 1 + \tan^2\left(\frac{\theta}{2}\right) = \sec^2\left(\frac{\theta}{2}\right)$,

so that $\qquad \dfrac{1}{1+t^2} = \cos^2\left(\frac{\theta}{2}\right)$.

Also $\qquad 1 - t^2 = \dfrac{\cos^2\left(\frac{\theta}{2}\right) - \sin^2\left(\frac{\theta}{2}\right)}{\cos^2\left(\frac{\theta}{2}\right)}$

and $\qquad 2t = 2\tan\left(\frac{\theta}{2}\right) = 2\dfrac{\sin\left(\frac{\theta}{2}\right)}{\cos\left(\frac{\theta}{2}\right)}$.

Therefore,

$$\frac{1-t^2}{1+t^2} = \cos^2\left(\frac{\theta}{2}\right) - \sin^2\left(\frac{\theta}{2}\right) = \cos(\theta) ,$$

$$\frac{2t}{1+t^2} = 2\sin\left(\frac{\theta}{2}\right)\cos\left(\frac{\theta}{2}\right) = \sin(\theta) ,$$

$$\frac{2t}{1-t^2} = \tan(\theta) .$$

These are very useful for solving linear trigonometric equations. For example, solve the equation

$$3\sin(\theta) + 4\cos(\theta) = 5 .$$

Using the above, we get

$$\frac{6t}{1+t^2} + \frac{4(1-t^2)}{1+t^2} = 5 ,$$

or

$$6t + 4 - 4t^2 = 5 + 5t^2$$

or

$$9t^2 - 6t + 1 = 0 .$$

What we generate is a quadratic equation in t. This may or may not have a "nice" solution, and if it has a solution, it may or may not lead to a "nice" angle θ.

In the example, we in fact have a perfect square, and thus, the solution is $t = \frac{1}{3}$, which generates an angle θ, but not in any "nice" form.

Problem 2.8. *Solve the following trigonometric equations:* (Solution on page 161.)

(1) $(a^2 - b^2)\sin(\theta) + 2ab\cos(\theta) = (a^2 + b^2)$.
(2) $\sin(\theta) + \cos(\theta) = 3$.
(3) $\sin(\theta) - \cos(\theta) = \sqrt{2}$.

The "half-angle" formulae are also used as substitutions in integration, to change some trigonometric integrals, with rational functions of the trigonometric function, into rational functions, which can often be integrated using partial fraction techniques.
If we have rational functions of **even** multiples of θ, then we can generate a similar substitution using

$$t = \tan(\theta) \ .$$

Then we have

$$1 + t^2 = 1 + \tan^2(\theta) = \sec^2(\theta) \ ,$$

so that $$\frac{1}{1+t^2} = \cos^2(\theta) \ .$$

Also $$1 - t^2 = \frac{\cos^2(\theta) - \sin^2(\theta)}{\cos^2(\theta)}$$

and $$2t = 2\tan(\theta) = 2\frac{\sin(\theta)}{\cos(\theta)} \ .$$

Therefore,

$$\frac{1-t^2}{1+t^2} = \cos^2(\theta) - \sin^2(\theta) = \cos(2\theta) \ ,$$

$$\frac{2t}{1+t^2} = 2\sin(\theta)\cos(\theta) = \sin(2\theta) \ ,$$

$$\frac{2t}{1-t^2} = \tan(2\theta) \ .$$

Problem 2.9. *Suppose that A, B, C are the angles of a triangle. Prove the following:* (Solution on page 162.)

(1) $\sin(A) + \sin(B) + \sin(C) = 4\cos(A/2)\cos(B/2)\cos(C/2)$;
(2) $\cos(A) + \cos(B) + \cos(C) = 1 + 4\sin(A/2)\sin(B/2)\sin(C/2)$;
(3) $\sin(2A) + \sin(2B) + \sin(2C) = 4\sin(A)\sin(B)\sin(C)$;

(4) $\cos(2A) + \cos(2B) + \cos(2C) = -\left(1 + 4\cos(A)\cos(B)\cos(C)\right)$;
(5) $\tan(A) + \tan(B) + \tan(C) = \tan(A)\tan(B)\tan(C)$;
(6) $\sin^2(A) + \sin^2(B) + \sin^2(C) = 2\left(1 + \cos(A)\cos(B)\cos(C)\right)$;
(7) $\cos^2(A) + \cos^2(B) + \cos^2(C) = 1 - 2\cos(A)\cos(B)\cos(C)$;
(8) $\cot(A)\cot(B) + \cot(B)\cot(C) + \cot(C)\cot(A) = 1$;
(9) $\cot(A/2) + \cot(B/2) + \cot(C/2) = \cot(A/2)\cot(B/2)\cot(C/2)$;
(10) $\left(\sin(A) + \sin(B) + \sin(C)\right) \times \left(-\sin(A) + \sin(B) + \sin(C)\right)$
$\times \left(\sin(A) - \sin(B) + \sin(C)\right) \times \left(\sin(A) + \sin(B) - \sin(C)\right)$
$= 4\sin^2(A)\sin^2(B)\sin^2(C)$.

2.8   The Sine Rule

We shall make use of the usual notation. That is, we use upper case letters
for the angles at the vertices of the triangle, and lower case letter for the
lengths of the opposite sides. It is standard in these cases to use A, B and
C for the three angles at the vertices of the triangle and, therefore, a, b and
c for the lengths of the opposite sides. We shall also, without confusion,
use A, B and C for the three vertices of the triangle. See Fig. 2.8.

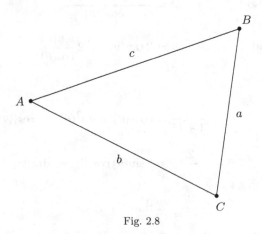

Fig. 2.8

We know that a circle can be drawn through three (non-collinear) points.
This circle is known as the **circumcircle** of the triangle. Its centre is
at the intersection of the perpendicular bisectors of the three sides. See
theorem 3.6 on page 55.

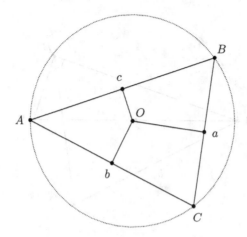

Fig. 2.9   The circle through A, B and C.

The Sine Rule is often stated as:

$$\frac{a}{\sin(A)} = \frac{b}{\sin(B)} = \frac{c}{\sin(C)} \ .$$

We shall prove it by showing that

$$\frac{a}{\sin(A)} = \frac{b}{\sin(B)} = \frac{c}{\sin(C)} = 2R \ ,$$

where R is the radius of the circumcircle of triangle ABC.
It is clearly sufficient to prove that

$$\frac{c}{\sin(C)} = 2R \ .$$

We shall therefore draw the diameter AP. See Fig. 2.10.
Assume, for the rest of this proof, that P and C lie on the same side of the chord AB. As an exercise, you should check that the result still holds on the other cases: when P and C lie on opposite sides of AB; or if P lies on AB.
We now see that $\angle ACB = \angle APB = C$. Since AP is a diameter, $\angle ABP$ is a right angle. Thus,

$$\frac{c}{\sin(C)} = \frac{c}{\sin(\angle APB)} = AP = 2R \ .$$

Problem 2.10.   *Complete the proof of the Sine Rule for the remaining cases.*

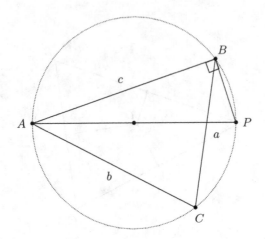

Fig. 2.10   Sine Rule

Problem 2.11.   *Find different proofs of the Sine Rule.*

Problem 2.12. *Prove* **Ptolemy's Theorem**; *that is, if ABCD is a cyclic quadrilateral, then AC.BD = AB.CD + AD.BC.* (Solution on page 167.)

2.9   The Cosine Rule

The Cosine Rule has three forms:

$$c^2 = a^2 + b^2 - 2ab\cos(C) \quad ;$$
$$b^2 = c^2 + a^2 - 2ca\cos(B) \quad ;$$
$$a^2 = b^2 + c^2 - 2bc\cos(A) \quad .$$

Clearly, if we obtain one form, we have the others.

We shall draw a perpendicular line from C to AB, intersecting AB at P. Denote the length of AP by c_a, the length of BP by c_b and the length of CP by h. See Fig. 2.11.

Assume for the rest of this proof that P lies between A and B. As an exercise, you should check that the result still holds in the other cases.

We have that $c = c_a + c_b$, so that

$$c^2 = c_a^2 + c_b^2 + 2c_a c_b$$
$$= a^2 - h^2 + b^2 - h^2 + 2c_a c_b$$

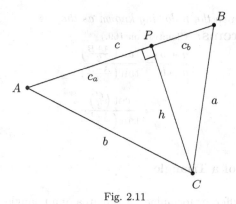

Fig. 2.11

so that it is now sufficient to prove that

$$h^2 = c_a c_b + ab\cos(C)$$

or
$$1 - \frac{c_a}{h}\frac{c_b}{h} = \frac{a}{h}\frac{b}{h}\cos(C)$$

or
$$1 - \cot(A)\cot(B) = \csc(A)\csc(B)\cos(C)$$

or
$$\sin(A)\sin(B) - \cos(A)\cos(B) = \cos(C) \ .$$

Since A, B and C are the angles of a triangle, we have that $A + B = \pi - C$, and the result follows at once.

This result is also known as the Extension of Pythagoras. If the angle C were a right angle, then we would get $c^2 = a^2 + b^2$, in other words, Pythagoras's Theorem.

Problem 2.13. *Complete the proof of the Cosine Rule for the remaining cases.*

Problem 2.14. *Find as many different proofs of the Cosine Rule as you can by searching through books.*

Problem 2.15. *Prove the following, known as*
Mollweide's Formulae: (Solution on page 168.)

$$\frac{a+b}{c} = \frac{\cos\left(\frac{A-B}{2}\right)}{\sin\left(\frac{C}{2}\right)}$$

$$\frac{a-b}{c} = \frac{\sin\left(\frac{A-B}{2}\right)}{\cos\left(\frac{C}{2}\right)}$$

Problem 2.16. *Prove the following known as the* **Tangent Theorems**: (Solution on 169.)

$$\frac{a+b}{a-b} = \frac{\tan\left(\frac{A+B}{2}\right)}{\tan\left(\frac{A-B}{2}\right)}$$

$$= \frac{\cot\left(\frac{C}{2}\right)}{\tan\left(\frac{A-B}{2}\right)}$$

2.10   The Area of a Triangle

There are several different formulae for the area of a triangle. The simplest is **"half of base times height"**. This is most easily seen by observing that a triangle is one half of a rectangle. We give one diagram (Fig. 2.12) to illustrate for an acute angled triangle:

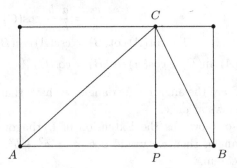

Fig. 2.12   Area of a triangle

Thus, $\Delta = \frac{1}{2}AB.CP$. We shall use this to get the formulae:

$$\Delta = \frac{1}{2}ab\sin(C) = \frac{1}{2}bc\sin(A) = \frac{1}{2}ca\sin(B) \ .$$

Let h be the length of the perpendicular from C to AB. See Fig. 2.13. Since $h = b\sin(A)$, we get

$$\Delta = \frac{1}{2}ch = \frac{1}{2}bc\sin(A) \ ,$$

and the others follow at once. Using the Sine Rule with this, we get yet another useful formula:

$$\Delta = \frac{abc}{4R} \ .$$

Problem 2.17. *Show that* $\Delta = 2R^2 \sin(A)\sin(B)\sin(C)$. (Solution on page 169.)

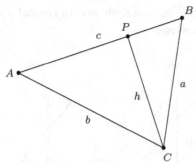

Fig. 2.13   Area of a triangle

2.11   Inverse Trigonometric Functions

We assume that you are familiar with the graphs of the trigonometric functions, and in particular, with the graphs of the sine, cosine and tangent functions.

The graph of the inverse of any function $y = f(x)$ is obtained by drawing the graph of $x = f(y)$. Of course, this is not necessarily the graph of a function of x. For example, the graph of $x = y^2$, the inverse of the squaring function $y = x^2$, is not the graph of a function of x.

2.11.1   *The Sine and Cosine Functions*

We will start here with the sine and cosine functions:

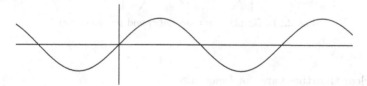

Fig. 2.14   Graph of $y = \sin(x)$.

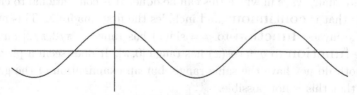

Fig. 2.15   Graph of $y = \cos(x)$.

We have drawn these here in true scale just to remind you of what that is.
Next we look at their inverses:

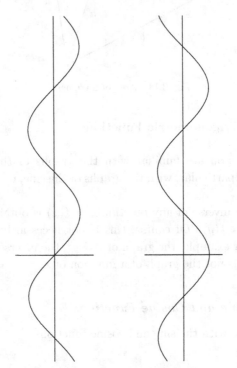

Fig. 2.16   Graphs of $y = \arcsin(x)$ and $y = \arccos(x)$

It is clear that these are not functions.

If we want to have functions, then we must restrict the **range**, so that
there is only one y–value for each x–value in the domain $[-1, 1]$. Clearly
there are many ways in which this can be done. It is conventional to choose
a range that is **continuous** and includes the first quadrant. This means
that the inverse **function** to $y = \sin(x)$ has range $[-\pi/2, \pi/2]$ and the
inverse **function** to $y = \cos(x)$ has range $[0, \pi]$. It does seem a pity that
they both do not have the same range, but an examination of the graphs
shows that this is not possible.

The inverse functions are denoted in one of several ways:

$$\sin^{-1}(x) \quad \text{Arc} \sin(x) \quad \text{Inv} \sin(x)$$
$$\cos^{-1}(x) \quad \text{Arc} \cos(x) \quad \text{Inv} \cos(x)$$

Capital letters are used at the beginning of "Arc sin" and "Inv sin" because "arc sin" and "inv sin" are used for the inverse **relations** as graphed in Fig. 2.26.

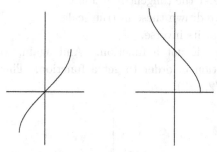

Fig. 2.17 Graphs of $y = \sin^{-1}(x)$ and $y = \cos^{-1}(x)$

Problem 2.18. *Obtain the following formulae:* (Solution on page 169.)

(1) $\cos^{-1}(x) + \sin^{-1}(x) = \frac{\pi}{2}$;

(2) $\sin^{-1}(-x) = - \sin^{-1}(x)$;

(3) $\cos^{-1}(-x) = \pi - \cos^{-1}(x)$;

(4) $\sin^{-1}(x) \pm \sin^{-1}(y) = \sin^{-1}\left(x\sqrt{1-y^2} \pm y\sqrt{1-x^2}\right)$
for $x^2 + y^2 \leq 1$;

(5) $\cos^{-1}(x) + \cos^{-1}(y) = \cos^{-1}\left(xy - \sqrt{(1-x^2)(1-y^2)}\right)$
Are there any restrictions on x and y?

(6) $\cos^{-1}(x) - \cos^{-1}(y) = - \cos^{-1}\left(xy + \sqrt{(1-x^2)(1-y^2)}\right)$
for $x \geq y$;

(7) $\cos^{-1}(x) - \cos^{-1}(y) = \cos^{-1}\left(xy + \sqrt{(1-x^2)(1-y^2)}\right)$
for $x < y$.

Problem 2.19. (Solution on page 173.) *Which of the following are true, which are false. Justify your answers!*

$$\cos\left(\cos^{-1}(x)\right) = x \qquad\qquad \cos^{-1}\left(\cos(\theta)\right) = \theta \ .$$

$$\sin\left(\sin^{-1}(x)\right) = x \qquad\qquad \sin^{-1}\left(\sin(\theta)\right) = \theta \ .$$

2.11.2   *The Tangent Function*

We shall now consider the tangent function.

Once again, we have drawn these in true scale.

Now we shall look at its inverse.

It is clear that this is not a function. And again, we shall make the appropriate restriction in order to get a function. This time we restrict that range to $(-\pi/2, \pi/2)$.

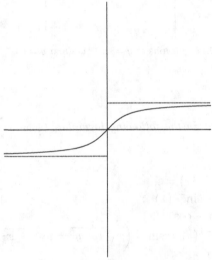

Fig. 2.18   Graph of $y = \tan^{-1}(x)$.

It is usual to define the inverse cotangent function from

$$\cot^{-1}(\theta) \ := \ \frac{\pi}{2} - \tan^{-1}(\theta) \ .$$

Problem 2.20. *Show that* $\tan^{-1}(x) \ = \ \sin^{-1}\left(\frac{x}{\sqrt{1+x^2}}\right)$. (Solution on page 173.)

Problem 2.21. *Obtain the following formulae:*

(1) $\tan^{-1}(-x) = -\tan^{-1}(x)$;

(2) $\cot^{-1}(-x) = \pi - \cot^{-1}(x)$;

(3) $\tan^{-1}(x) + \tan^{-1}(y) = \tan^{-1}\left(\frac{x+y}{1-xy}\right)$

    for $xy < 1$;

(4) $\tan^{-1}(x) - \tan^{-1}(y) = \tan^{-1}\left(\frac{x-y}{1+xy}\right)$

    for $xy > -1$;

(5) $\cot^{-1}(x) + \cot^{-1}(y) = \cot^{-1}\left(\frac{xy-1}{x+y}\right)$

    for $x + y \neq 0$;

(6) $\cot^{-1}(x) - \cot^{-1}(y) = \cot^{-1}\left(\frac{xy+1}{y-x}\right)$

    for $x \neq y$. (Solution on page 173.)

Problem 2.22. *Which of the following are true, which are false. Justify your answers!*

$$\tan\left(\tan^{-1}(x)\right) = x \qquad\qquad \tan^{-1}(\tan(\theta)) = \theta .$$

(Solution on page 175.)

2.12 De Moivre's Theorem

This section assumes knowledge of basic complex numbers. We denote the imaginary unit by i, so that $i^2 = -1$.

We shall combine the sine and cosine functions using i and, for notational ease, define

$$\mathbf{cis}(\theta) = \cos(\theta) + i\sin(\theta) .$$

The function **cis** is therefore a periodic function, with period 2π.

Using basic trigonometric identities, we can easily obtain that:

$$\mathbf{cis}(\theta)\,\mathbf{cis}(\phi) = \big(\cos(\theta) + i\sin(\theta)\big)\big(\cos(\phi) + i\sin(\phi)\big)$$

$$= \big(\cos(\theta)\cos(\phi) - \sin(\theta)\sin(\phi)\big)$$
$$+ i\big(\sin(\theta)\cos(\phi) + \cos(\theta)\sin(\phi)\big)$$

$$= \big(\cos(\theta + \phi) + i\sin(\theta + \phi)\big)$$
$$= \mathbf{cis}(\theta + \phi) .$$

Problem 2.23. *Show that* $\mathbf{cis}(\theta)\,\mathbf{cis}(\phi)\,\mathbf{cis}(\psi) = \mathbf{cis}(\theta + \phi + \psi)$. (Solution on page 175.)

   Show that, for any integer n,

(1) $\prod_{k=1}^{n} \mathbf{cis}(\theta_k) = \mathbf{cis}\left(\prod_{k=1}^{n} \theta_k\right)$;
(2) $\left(\mathbf{cis}(\theta)\right)^n = \mathbf{cis}(n\theta)$.   *This is* **De Moivre's** *Theorem.*

Problem 2.24. *Use De Moivre's theorem to prove that*

$$\left(\mathbf{cis}(\theta)\right)^{1/n} = \mathbf{cis}\left(\frac{\theta}{n}\right)$$

for every natural number n.
Deduce De Moivre's theorem for every rational number n. (Solution on page 176.)

2.12.1   *Roots of Unity and Euler's Formula*

De Moivre's theorem is useful for extracting roots of complex numbers (in polar form), but care must be taken since the representation of a complex number in polar form is not unique, for it is easy to see that $\mathbf{cis}(\theta + 2k\pi) = \mathbf{cis}(\theta)$ for every integer k. This gives that

$$\left(\mathbf{cis}(\theta)\right)^{1/n} = \mathbf{cis}\left(\frac{\theta + k\pi}{n}\right) ,$$

where $k \in \{0, 1, \ldots, (n-1)\}$.

From this, we get the **roots of unity** by setting $\theta = 0$. We see that these are given as points evenly distributed around the unit circle.

The use of **Euler's formula**, $\mathbf{cis}(\theta) = e^{i\theta}$, although difficult to justify with the knowledge available at this stage, makes calculations much more straight forward.

With this form, we see that the exponential function is a periodic function, but that the period is an **imaginary** number, $2\pi i$.

We can make use of De Moivre's theorem to obtain formulae for the sine and cosine of integer multiples of angles. For example,

$$\mathbf{cis}(3\theta) = \left(\mathbf{cis}(\theta)\right)^3$$

$$= \left(\cos^3(\theta) - 3\cos(\theta)\sin^2(\theta)\right)$$
$$+ i\left(3\cos^2(\theta)\sin(\theta) - \sin^3(\theta)\right) ,$$

from which we obtain

$$\cos(3\theta) = \cos^3(\theta) - 3\cos(\theta)\sin^2(\theta)$$

$$\sin(3\theta) = 3\cos^2(\theta)\sin(\theta) - \sin^3(\theta) \ .$$

Problem 2.25. *Prove the following formulae:* (Solution on page 176.)

(1) $\sin(3\theta) = 3\sin(\theta) - 4\sin^3(\theta)$;
(2) $\cos(3\theta) = 4\cos^3(\theta) - 3\cos(\theta)$;
(3) $\sin(4\theta) = 8\sin(\theta)\cos^3(\theta) - 4\sin(\theta)\cos(\theta)$;
(4) $\cos(5\theta) = 16\cos^5(\theta) - 20\cos^3(\theta) + 5\cos(\theta)$;
(5) $\tan(3\theta) = \frac{3\tan(\theta) - \tan^3(\theta)}{1 - 3\tan^2(\theta)}$;
(6) $\cot(4\theta) = \frac{\cot^4(\theta) - 6\cot^2(\theta) + 1}{4\cot^3(\theta) - 4\cot(\theta)}$.

Problem 2.26. *Find general formulae for* $\sin(n\theta)$ *and* $\cos(n\theta)$. (*You will need to make use of binomial coefficients*). (Solution on page 178.)

Problem 2.27. *Prove the following formulae:* (Solution on page 179.)

(1) $\sin^3(\theta) = \frac{1}{4}\left(3\sin(\theta) - \sin(3\theta)\right)$;

(2) $\cos^3(\theta) = \frac{1}{4}\left(3\cos(\theta) + \cos(3\theta)\right)$;

(3) $\sin^4(\theta) = \frac{1}{8}\left(\cos(4\theta) - 4\cos(2\theta) + 3\right)$;

(4) $\cos^5(\theta) = \frac{1}{16}\left(10\cos(\theta) + 5\cos(3\theta) + \cos(5\theta)\right)$.

Since $e^{i\theta} = \cos(\theta) + i\sin(\theta)$, we also have $e^{-i\theta} = \cos(\theta) - i\sin(\theta)$. This now leads to the formulae

$$\cos(\theta) = \frac{e^{i\theta} + e^{-i\theta}}{2} \ ,$$

$$\sin(\theta) = \frac{e^{i\theta} - e^{-i\theta}}{2i} \ ,$$

$$\tan(\theta) = \frac{-i\left(e^{i\theta} - e^{-i\theta}\right)}{e^{i\theta} + e^{-i\theta}} \ .$$

The sine and cosine functions are useful to give parametric equations for a circle (or ellipse), and thus, are known as **circular** functions.

For the circle given by $x^2 + y^2 = r^2$, the pair of equations

$$x = r\,\cos(\theta)$$
$$y = r\,\sin(\theta)$$

with $0 \le \theta \le 2\pi$, describes the circle once round.

For the general circle given by $(x-\alpha)^2 + (y-\beta)^2 = r^2$, the pair of equations is

$$x = \alpha + r\,\cos(\theta)$$
$$y = \beta + r\,\sin(\theta)$$

with $0 \le \theta \le 2\pi$.

Problem 2.28. *Find the parametric equations for the ellipse in standard position*

$$\frac{x^2}{a^2} + \frac{y^2}{b^2} = 1 \ .$$

(Solution on page 180.)

Finally, we can use De Moivre's theorem to calculate sums like:

$$S(\theta) = \sum_{k=\lambda+1}^{\mu} \sin(k\theta)$$

and

$$C(\theta) = \sum_{k=\lambda+1}^{\mu} \cos(k\theta) \ .$$

We do this by first calculating

$$\sum_{\lambda+1}^{\mu} e^{ik\theta} \ ,$$

a geometric series, and then taking real and imaginary parts.
We note that

$$e^{i\theta} - 1 = e^{i\theta/2}\left(e^{i\theta/2} - e^{-i\theta/2}\right) = 2i\,e^{i\theta/2}\sin(\theta/2) \ .$$

Therefore, we have

$$\sum_{k=\lambda+1}^{\mu} e^{ik\theta} = \frac{e^{i(\mu+1)\theta/2} - e^{i(\lambda+1)\theta/2}}{e^{i\theta/2} - 1}$$

$$= \frac{e^{i(\mu+1)\theta/2} - e^{i(\lambda+1)\theta/2}}{2i\, e^{i\theta/2} \sin(\theta/2)}$$

$$= -i\frac{e^{i(\mu+1/2)\theta/2} - e^{i(\lambda+1/2)\theta/2}}{2\sin(\theta/2)}$$

$$= \left\{ \frac{\cos\left((\mu+1/2)\theta\right) - \cos\left((\lambda+1/2)\theta\right)}{2\sin(\theta/2)} \right\}$$
$$+ i \left\{ \frac{\cos\left((\lambda+1/2)\theta\right) - \cos\left((\mu+1/2)\theta\right)}{2\sin(\theta/2)} \right\}$$

$$= \left(\frac{\cos\left\{(\mu+\lambda+1)\theta/2\right\} \sin\left\{(\mu-\lambda)\theta/2\right\}}{\sin(\theta/2)} \right)$$
$$+ i \left(\frac{\sin\left\{(\mu+\lambda+1)\theta/2\right\} \sin\left\{(\mu-\lambda)\theta/2\right\}}{\sin(\theta/2)} \right).$$

Thus, we have

$$\sum_{k=\lambda+1}^{\mu} \cos(k\theta) = \left(\frac{\cos\left\{(\mu+\lambda+1)\theta/2\right\} \sin\left\{(\mu-\lambda)\theta/2\right\}}{\sin(\theta/2)} \right),$$

$$\sum_{k=\lambda+1}^{\mu} \sin(k\theta) = \left(\frac{\sin\left\{(\mu+\lambda+1)\theta/2\right\} \sin\left\{(\mu-\lambda)\theta/2\right\}}{\sin(\theta/2)} \right).$$

Problem 2.29. *Prove that, for* $0 < \theta_k < \pi$,

$$\left| \sin\left(\sum_{k=1}^{\mu} \theta_k \right) \right| < \sum_{k=1}^{\mu} \sin(\theta_k) \ .$$

(Solution on page 180.)

Problem 2.30. *If* $S_c = \sum_{k=\lambda+1}^{\mu} \cos(k\theta)$ *and* $S_s = \sum_{k=\lambda+1}^{\mu} \sin(k\theta)$, *show that, for* $0 < \theta < 2\pi$,

$$\sqrt{S_c^2 + S_s^2} \ \le \ \frac{1}{\sin(\theta/2)} \ .$$

(Solution on page 181.)

2.13   Hyperbolic Functions

The **Hyperbolic** functions are defined from the exponential function as follows

$$\cosh(x) = \frac{e^x + e^{-x}}{2}$$
$$\sinh(x) = \frac{e^x - e^{-x}}{2}$$

with $\tanh(x)$, $\coth(x)$, **sech**(x) and **csch**(x) defined analogously to the way that $\tan(\theta)$, $\cot(\theta)$, $\sec(\theta)$ and $\csc(\theta)$ are obtained from $\cos(\theta)$ and $\sin(\theta)$. The following two formulae are immediate:

$$\cosh(x) + \sinh(x) = e^x \ ,$$
$$\cosh(x) - \sinh(x) = e^{-x} \ .$$

It is now easy to show that

$$\cosh^2(x) - \sinh^2(x) \ = \ 1 \ .$$

Thus, it follows that the pair of equations

$$x = \cosh(x)$$
$$y = \sinh(x)$$

are parametric equations for the hyperbola $x^2 - y^2 \ = \ 1$. Hence, the name **hyperbolic** functions.

Problem 2.31. *Show the following properties for all $x \in R$:* (Solution on page 181.)

(1) $\sinh(x) \in (-\infty, \infty)$;
(2) $\cosh(x) \in [1, \infty)$;
(3) $\tanh(x) \in (-1, 1)$.

Problem 2.32. *Show that* (Solution on page 182.)

(1) $\coth(x) \in (1, \infty)$ *for* $x \in (0, \infty)$;
(2) $\coth(x) \in (-\infty, -1)$ *for* $x \in (-\infty, 0)$;
(3) $\lim\limits_{x \to \infty} \tanh(x) = \lim\limits_{x \to \infty} \coth(x) = 1$;
(4) $\lim\limits_{x \to -\infty} \tanh(x) = \lim\limits_{x \to -\infty} \coth(x) = -1$.

Problem 2.33. *Show that*

(1) $\sinh(2x) = 2\sinh(x)\cosh(x)$;
(2) $\cosh(2x) = \cosh^2(x) + \sinh^2(x)$;
(3) $\tanh(2x) = \frac{2\tanh(x)}{1 + \tanh^2(x)}$.

Find two other expressions for $\cosh(2x)$. (Solution on page 182.)

There are many formulae for hyperbolic functions that are very similar to those for trigonometric functions. We could easily write down a large selection here as problems. But that would be a waste of your effort, for the use of the following relations makes the translation between hyperbolics and trigonometrics quite easy:

$$\cosh(i\theta) = \cos(\theta) , \qquad \cos(i\theta) = \cosh(\theta) ,$$

$$\sinh(i\theta) = i\sin(\theta) , \qquad \sin(i\theta) = i\sinh(\theta) .$$

Problem 2.34. *Translate all the trigonometric identities in the exercises above to hyperbolic identities.* (Solution on page 183.)

2.14 Inverse Hyperbolic Functions

We start this section with the graphs of the hyperbolic functions.
Note that, as $x \to \infty$ both $\sinh(x)$ and $\cosh(x)$ are asymptotic to $\frac{e^x}{2}$ and $\tanh(x)$ is asymptotic to 1.
We will look at the inverses. Clearly, both $\sinh(x)$ and $\tanh(x)$ have inverse functions, whereas $\cosh(x)$ needs to have a restriction placed on the range. It is usual to insist that $x \geq 0$ in this case.
This leads to the definitions:

$y = \sinh^{-1}(x)$ if and only if $x = \sinh(y)$.
$y = \cosh^{-1}(x)$ if and only is $x = \cosh(y)$ **and** $x \geq 1$.
$y = \tanh^{-1}(x)$ if and only if $x = \tanh(y)$.

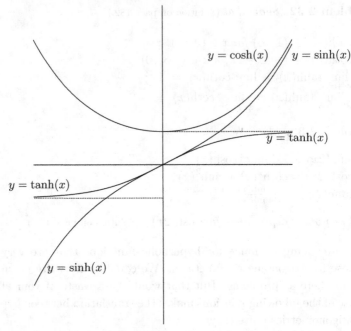

Fig. 2.19   Graphs of $y = \sinh(x)$, $y = \cosh(x)$ and $y = \tanh(x)$.

We see that the domains are all different:

Function	Domain	Range
$\sinh^{-1}$	$(-\infty, \infty)$	$(-\infty, \infty)$
$\cosh^{-1}$	$[1, \infty)$	$[0, \infty)$
$\tanh^{-1}$	$[-1, 1]$	$(-\infty, \infty)$

It is also conventional to define $\coth^{-1}(x)$ by the relation

$$\coth^{-1}(x) = \tanh^{-1}(1/x) .$$

Problem 2.35. *Obtain the following formulae:* (Solution on page 183.)

(1) $\sinh^{-1}(x) = \cosh^{-1}\left(\sqrt{x^2 + 1}\right)$ *for* $x \geq 0$;
(2) $\sinh^{-1}(x) = -\cosh^{-1}\left(\sqrt{x^2 + 1}\right)$ *for* $x < 0$;
(3) $\sinh^{-1}(x) = \tanh^{-1}\left(\frac{x}{\sqrt{x^2+1}}\right)$;
(4) $\sinh^{-1}(x) = \coth^{-1}\left(\frac{\sqrt{x^2+1}}{x}\right)$;
(5) $\cosh^{-1}(x) = \left|\sinh^{-1}\left(\sqrt{x^2 - 1}\right)\right|$;
(6) $\cosh^{-1}(x) = \left|\tanh^{-1}\left(\frac{\sqrt{x^2-1}}{x}\right)\right|$;

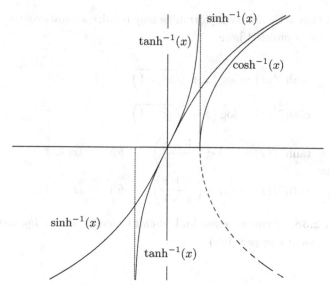

Fig. 2.20   Graphs of $y = \sinh^{-1}(x)$, $y = \cosh^{-1}(x)$ and $y = \tanh^{-1}(x)$.

(7) $\cosh^{-1}(x) = \left| \coth^{-1}\left(\frac{x}{\sqrt{x^2-1}} \right) \right|$;

(8) $\tanh^{-1}(x) = \sinh^{-1}\left(\frac{x}{\sqrt{1-x^2}} \right)$;

(9) $\tanh^{-1}(x) = \cosh^{-1}\left(\frac{1}{\sqrt{1-x^2}} \right)$ *for* $x > 0$;

(10) $\tanh^{-1}(x) = -\cosh^{-1}\left(\frac{1}{\sqrt{1-x^2}} \right)$ *for* $x < 0$.

Problem 2.36. *Obtain the following formulae:* (Solution on page 184.)

(1) $\sinh^{-1}(x) \pm \sinh^{-1}(y) = \sinh^{-1}\left(x\sqrt{1+y^2} \pm y\sqrt{1+x^2} \right)$;

(2) $\cosh^{-1}(x) \pm \cosh^{-1}(y) = \cosh^{-1}\left(xy \pm \sqrt{(1+x^2)(1+y^2)} \right)$;

(3) $\tanh^{-1}(x) \pm \tanh^{-1}(y) = \tanh^{-1}\left(\frac{x \pm y}{1 \pm xy} \right)$.

Problem 2.37. *Obtain the following formulae:* (Solution on page 184.)

(1) $\sinh^{-1}(i\theta) = i\sin^{-1}(\theta)$;

(2) $\cosh^{-1}(i\theta) = i\cos^{-1}(\theta)$;

(3) $\tanh^{-1}(i\theta) = i\tan^{-1}(\theta)$.

The derivation of the following formulae may require a knowledge of integral calculus, and is omitted here.

$$\sinh^{-1}(x) = \log\left(x + \sqrt{x^2 + 1}\right) \; ;$$

$$\cosh^{-1}(x) = \log\left(x + \sqrt{x^2 - 1}\right) \; ;$$

$$\tanh^{-1}(x) = \frac{1}{2}\log\left(\frac{1+x}{1-x}\right) \qquad \text{for} \qquad |x| < 1 \; ;$$

$$\coth^{-1}(x) = \frac{1}{2}\log\left(\frac{x+1}{x-1}\right) \qquad \text{for} \qquad |x| > 1 \; .$$

Problem 2.38. *If you have the background Calculus, derive the above four formulae.* (Solution on page 185.)

Chapter 3

Concurrency and Collinearity

In this chapter, we look at the standard results about concurrency and collinearity. We shall show that the following are concurrent:

The medians of a triangle.

The altitudes of a triangle.

The perpendicular bisectors of the sides of a triangle.

The internal bisectors of the angles of a triangle.

An internal bisector of an angle of a triangle with the external bisectors of the other two angles.

We shall begin with a result on collinearity, **Menelaus' Theorem**.

3.1 Menelaus' Theorem

Menelaus was a Greek Mathematician, who lived in Alexandria about 100 A.D. He is known as the Father of Spherical Trigonometry.

This is a result about the intersection of a line with the sides of a triangle. To avoid unnecessary problems at this stage, the line is taken so that it is not parallel to any of the line segments that form the triangle, and so that it does not pass through any vertex.

It is clear that this line cannot intersect all three of the line segments: either it intersects two of the line segments, or it cuts none of them.

Therefore, we shall allow a line segment of the triangle to be extended whenever necessary, to meet the "other" line. For simplicity, we shall only draw one of the two pictures that may occur.

We shall name the triangle $\triangle ABC$, the line l, and the points of intersection of l with the sides AB, BC and CA, as L, M and N, respectively.

Fig. 3.1   A triangle and a straight line

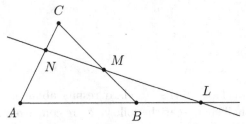

Fig. 3.2   A straight line intersecting a side of a triangle

Theorem 3.1. Menelaus' Theorem

If three points L, M and N lying on the sides (extended as necessary) AB, BC and CA of triangle $\triangle ABC$, are collinear, then we have

$$\frac{AL}{LB} \cdot \frac{BM}{MC} \cdot \frac{CN}{NA} = -1 \ .$$

The answer is given as -1 since we consider **directed** line segments. If we were to consider only **lengths**, then the answer would be 1.

Proof.

Draw perpendiculars from A, B and C to l, to meet l at P, Q and R, respectively.

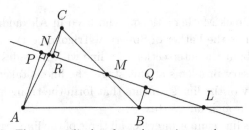

Fig. 3.3   The perpendiculars from the vertices to the transverse line

This leads to some similar triangles:

$$\triangle APL \sim \triangle BQL \ ,$$

$$\triangle BQM \sim \triangle CRM \ ,$$

$$\triangle CRN \sim \triangle APN \ .$$

(This is so because $AP \parallel BQ \parallel CR$ – all are perpendicular to l.)
From these similar triangles, we now get that

$$\frac{AL}{LB} = \frac{AP}{QB} ,$$

$$\frac{BM}{MC} = \frac{BQ}{RC} ,$$

$$\frac{CN}{NA} = \frac{CR}{PA} ,$$

and we now get the result.

This completes the proof of Menelaus' Theorem.

An alternate way of writing the result is

$$AL \cdot BM \cdot CN = BL \cdot CM \cdot AN .$$

Problem 3.1. *Investigate* **degenerate** *cases of Menelaus' Theorem.*
For example, what is the case if l passes through a vertex?
Can we interpret the theorem if l is parallel to a side? (Solution on page 185.)

Theorem 3.2. Converse of Menelaus' Theorem

If three points L, M and N lying on the sides (extended as necessary) AB,
BC and CA of triangle $\triangle ABC$, satisfy

$$\frac{AL}{LB} \cdot \frac{BM}{MC} \cdot \frac{CN}{NA} = -1 ,$$

then the three points L, M and N are collinear.

Proof.

Let the line LM meet AC at N'. Then, by Menelaus' theorem, we have

$$\frac{AL}{LB} \cdot \frac{BM}{MC} \cdot \frac{CN'}{N'A} = -1 .$$

Comparing this with

$$\frac{AL}{LB} \cdot \frac{BM}{MC} \cdot \frac{CN}{NA} = -1 ,$$

we are able to deduce that $\frac{CN}{NA} = \frac{CN'}{N'A}$. It follows that $NA = N'A$, so that
N and N' are the same point.

Problem 3.2. *Verify that N and N' are the same point.*

3.2   Ceva's Theorem

Ceva was an Italian Mathematician who lived from about 1647 to 1734. He
discovered this theorem when he was about 30 years old.

Let $\triangle ABC$ be a triangle and P a point, not on the triangle.

Draw the line segments AP, BP and CP, and extend them if necessary to
meet the sides of the triangle at D, E and F, respectively.

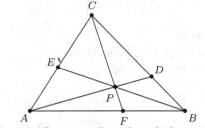

Fig. 3.4   Concurrent lines through the vertices

Theorem 3.3. Ceva's Theorem

*If three lines AD, BE and CF on the vertices A, B and C of triangle
$\triangle ABC$, with D on BC, E on CA and F on AB, are concurrent, then we
have*

$$\frac{AF}{FB} \cdot \frac{BD}{DC} \cdot \frac{CE}{EA} = 1 \ .$$

Proof.

We prove Ceva's Theorem by using Menelaus' Theorem on two different
triangles: first to triangle $\triangle ABD$ and line CF, and then to triangle $\triangle ACD$
and line BE.

Let P be the point where AD, BE and CF are concurrent.

For $\triangle ABD$ and line CF, we have

$$\frac{AP}{PD} \cdot \frac{DC}{CB} \cdot \frac{BF}{FA} = -1 \ ,$$

and for $\triangle ACD$ and line BE, we have

$$\frac{AE}{EC} \cdot \frac{CB}{BD} \cdot \frac{DP}{PA} = -1 \ .$$

By multiplying these equations together, we get

$$\frac{AP}{PD} \cdot \frac{DC}{CB} \cdot \frac{BF}{FA} \cdot \frac{AE}{EC} \cdot \frac{CB}{BD} \cdot \frac{DP}{PA} = 1 \ ,$$

which may be re-written as

$$\frac{AF}{FB} \cdot \frac{BD}{DC} \cdot \frac{CE}{EA} = 1 \ .$$

This completes the proof of Ceva's Theorem.

Problem 3.3. *Prove the converse of Ceva's Theorem: that is, if points D, E and F on sides BC, CA and AB, respectively, of triangle △ABC satisfy*

$$\frac{AF}{FB} \cdot \frac{BD}{DC} \cdot \frac{CE}{EA} = 1 \ ,$$

then the lines AD, BE and CF are concurrent. (Solution on page 186.)

3.3 Particular Concurrency Results

Here we consider

- Medians,
- Altitudes,
- Perpendicular Bisectors of the Sides,
- Interior Angle Bisectors,
- Exterior Angle Bisectors.

3.3.1 *Medians*

Definition 3.1. A **median** is a line joining a vertex of a triangle to the **mid-point** of the opposite side.

Theorem 3.4. *The medians of a triangle are concurrent.*

Proof.
We make use of the converse of Ceva's Theorem.
It is conventional to denote the mid-points of the sides AB, BC and CA by N, L and M, respectively, and the point at which they concur by G. This point is known as the **centroid**.
We use the notation in the diagram. Using the fact that $AN = NB$, $BL = LC$ and $CM = MA$, we obtain that

$$\frac{AN}{NB} \cdot \frac{BL}{LC} \cdot \frac{CM}{MA} = 1 \ .$$

By the converse of Ceva's Theorem, the proof is complete.

Problem 3.4. *In Fig. 3.5, show that $\frac{AG}{GL} = \frac{BG}{GM} = \frac{CG}{GN} = 2$.*

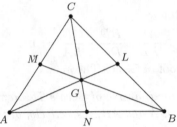

Fig. 3.5   The medians of a triangle

Problem 3.5. *In Fig. 3.5, show that*

(1) $ML \parallel AB$,
(2) $ML = \frac{AB}{2}$.
    Find similar results for the other sides of the triangle.

3.3.2   Altitudes

Definition 3.2. An **altitude** is the line on a vertex, perpendicular to the opposite side.

Theorem 3.5. *The altitudes of a triangle are concurrent.*

It is conventional to denote the feet of the altitudes from A, B and C by D, E and F, respectively, and the point at which they concur by H. This point is known as the **orthocentre**.
Proof.
We give the proof in the case when no angle is greater than $\pi/2$.
Let A, B and C be the vertices of the triangle and let AD and BE be altitudes. Suppose that they intersect at H. Draw the line segment CH and let it meet (extended if necessary) the side AB at F. (We are not assuming that CF is an altitude.)

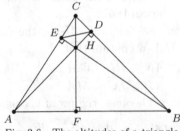

Fig. 3.6   The altitudes of a triangle

Since $\angle AEB = \angle ADB = \pi/2$, we see that quadrilateral $AEDB$ is cyclic. From this we deduce that

$$\angle DAB = \angle DEB \ ,$$

or, equivalently,

$$\angle DAF = \angle DEH \ .$$

Since $\angle CEH = \angle CDH = \pi/2$, we also see that quadrilateral $CEHD$ is cyclic. From this we deduce that

$$\angle DEH = \angle DCH \ ,$$

or, equivalently,

$$\angle DEH = \angle DCF \ .$$

This gives that

$$\angle DAF = \angle DCF$$

so that quadrilateral $CDFA$ is cyclic. And we deduce that

$$\angle CFA = \angle CDA = \pi/2 \ ,$$

so that AF is perpendicular to AB. That is, AF is the third altitude of $\triangle ABC$. Thus, the proof is complete.

Problem 3.6. *Give the proof in the case where one of the angles is equal to $\pi/2$.*

Problem 3.7. *Give the proof in the case where one of the angles is greater than $\pi/2$.*

3.3.3 *Perpendicular Bisectors of the Sides*

Definition 3.3. A **Perpendicular Bisector** of a side of a triangle is the line on the mid-point of the side, perpendicular to that side.

Theorem 3.6. *The perpendicular bisectors of the sides of a triangle are concurrent.*

The mid-points of the sides AB, BC and CA are denoted, as before, by N, L and M, respectively. The point at which the perpendicular bisectors concur is denoted conventionally by O. This point is known as the **centre** or **circumcentre**, for it is the centre of the circle that passes through A, B and C.

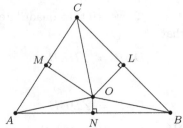

Fig. 3.7 The perpendicular bisectors of the sides of a triangle.

Proof.

Suppose that LO and MO are the perpendicular bisectors of BC and CA, respectively. Let N be the mid-point of AB. We are not assuming that ON is perpendicular to AB.

Since $BL = LC$, $\angle BLO = \angle CLO = \pi/2$ and $LO = LO$, we see that $\triangle BOL$ and $\triangle COL$ are congruent. This gives that $BO = CO$.

In a similar way, we see that $\triangle COM$ and $\triangle AOM$ are congruent. This gives that $CO = AO$. Thus, $AO = BO$.

We now have that $AO = BO$, $AN = BN$ and $ON = ON$. Thus, the triangles $\triangle AON$ and $\triangle BON$ are congruent. From this we deduce that $\angle ANO = \angle BNO$, and since $\angle ANB = \pi$, we obtain that ON is the perpendicular bisector of AB. The proof is now complete.

Since we have shown that $OA = OB = OC$, we see that O is the centre of the circle passing through A, B and C. This circle is known as the **circumcircle**. It is conventional to denote the radius of this circle by R.

3.3.4 *Interior Angle Bisectors*

Definition 3.4. An **interior angle bisector** is the line on a vertex of a triangle that bisects the angle at that vertex, interior to the triangle.

This is usually just called the **angle bisector**.

Theorem 3.7. *The interior angle bisectors of a triangle are concurrent.*

Suppose that AU, BV and CW are the internal bisectors of angles $\angle CAB$, $\angle ABC$ and $\angle CBA$, respectively, where U, V and W lie on the sides BC, CA and AB, respectively. The point at which the internal bisectors concur is conventionally denoted by I. This point is known as the **in-centre** for it is the centre of the circle which is tangent to the three sides of the triangle.

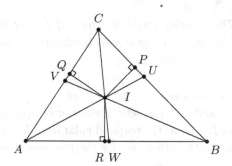

Fig. 3.8 The interior angle bisectors of a triangle.

Proof.

Suppose that AU and BV are the internal bisectors of angles $\angle CAB$ and ABC with U and V on BC and CA, respectively. Suppose that they meet at I. To help, draw the line segment IC. We are not assuming that IC is the internal bisector of angle $\angle BAC$. Suppose that IC meets AB at W.

Draw perpendiculars for I to BC, CA and AB, meeting those sides at P, Q and R, respectively.

Since $\angle QAI = \angle RAI$, $\angle IRA = \angle IQA = \pi/2$ and $IA = IA$, we see that triangles $\triangle QAI$ and $\triangle RAI$ are congruent. This gives that $IQ = IR$.

In a similar way, we see that triangles $\triangle RBI$ and $\triangle PBI$ are congruent, so that $IR = IP$.

We now have that $\angle CPI = \angle CQI = \pi/2$, $IP = IQ$ and $IC = IC$. Thus, triangles $\triangle CQI$ and $\triangle CPI$ are congruent.

This implies that $\angle ICP = \angle ICQ$, so that IC is the internal bisector of $\angle ACB$. The proof is now complete.

Since we have shown that $IP = IQ = IR$, we see that the circle, radius IP is tangent to each of the sides of the circle. This is called the **in-circle**. It is conventional to denote the radius of this circle by r.

Problem 3.8. *Show that if any two of the centroid, the orthocentre, the circumcentre, or the incentre, coincide, then the triangle must be equilateral. (Hence, all four coincide.)*

3.3.5 Exterior Angle Bisectors

Definition 3.5. An **exterior angle bisector** is the line on a vertex of a triangle that bisects the angle at that vertex, exterior to the triangle.

This line is **perpendicular** to the interior angle bisector.

Theorem 3.8. *The exterior angle bisectors of two of the angles at the vertices of a triangle are concurrent with the interior angle bisector at the third vertex.*

The proof of this theorem is left as an exercise.

Hint: Extend CA and CB, and let the exterior bisectors of angles $\angle CAB$ and $\angle CBA$ meet at I_C. Draw the perpendicular from I_C to AB and proceed in a manner similar to the proof give for the previous theorem.

3.3.6   The Gergonne Point

Here we consider the in-circle and the points where it is tangent to the sides.

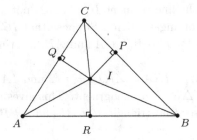

Fig. 3.9   The Gergonne Point.

In the nineteenth century, the French mathematician, J.D. Gergonne, discovered the following result.

Theorem 3.9. *The three lines AP, BQ and CR are concurrent.*

Proof.
Since $AQ = AR$, $BR = BP$ and $CP = CQ$, we have that

$$\frac{AR}{RB} \cdot \frac{BP}{PC} \cdot \frac{CQ}{QA} = 1 \ .$$

The concurrency of AP, BQ and CR now follows from the converse of Ceva's Theorem. Therefore, the proof is complete.

The point at which the three line segments, AP, BQ and CR concur is known as the **Gergonne Point**.

3.4 The Wallace-Simson Line

William Wallace was a Scottish mathematician who lived from 1768 to 1843. His publication, in 1799, of this concept, predates Simson's. Robert Simson was an English mathematician who lived from 1687 to 1768. He is not to be confused with Thomas Simpson (1710-1761) after whom is named **Simpson's Rule** in numerical integration.

Here we are interested in a point on the circumcircle of a triangle, and the perpendiculars from that point to the side of the triangle.

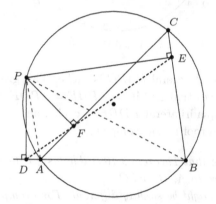

Fig. 3.10   The Wallace-Simson Line

Theorem 3.10. *Suppose that P is a point on the circumcircle of triangle $\triangle ABC$, not coincident with any vertex.*

Suppose that PD, PE and PF are perpendicular to AB, BC and CA, respectively, with D on AB, E on BC and F on CA (the line segments being extended as necessary).

Then D, E and F are collinear.

The line on D, E and F is called the **Wallace-Simson Line**.

Proof.

In the above diagram, we shall assume only that PD is perpendicular to AB and that PF is perpendicular to CA. We draw the line DF and let it (extended as necessary) intersect CB at E. Draw the line segment PE. We shall show that PE is perpendicular to BC.

To help, we draw the line segments PA and PB.

First, since $\angle PDA = \angle PFA = \pi/2$, we see that quadrilateral $PDAF$ is cyclic. This gives that $\angle PDF = \angle PAF$. Equivalently, $\angle PDE = \angle PAC$.

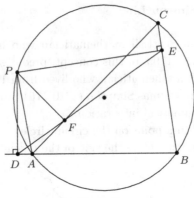

Fig. 3.11

Next, we see that quadrilateral $PABC$ is cyclic. This gives that $\angle PAC = \angle PBC$. And this gives that $\angle PAC = \angle PBE = \angle PDE$.

We now have that quadrilateral $PDBE$ is cyclic, and it now follows that $\angle PEB = \pi/2$. The proof is now complete.

Problem 3.9. *Show that Theorem 3.10 still holds if P takes other positions on the circle, relative to A, B and C.*

Note that the proof might be slightly different. For example, what if D lies between A and B? What if P and C are of "different" arcs? Try to consider all possible cases.

3.5   The Nine Point Circle

Consider triangle $\triangle ABC$. Let L, M and N be the mid-points of BC, CA and AB, respectively. Let AD, BE and CF be the altitudes.

Theorem 3.11.   The Six Point Circle Theorem

For triangle $\triangle ABC$, the mid-points of the sides, L, M and N, and the feet of the altitudes D, E and F, all lie on a circle.

Proof.

We give the proof in the case where all angles are less than $\pi/2$.

To help, draw the line segments ML, MF and LN. We shall also indicate the circle that passes through the six points.

First we recall that any three points lie on a circle. Thus, we take the circle that passes through the three mid-points, L, M and N. We shall show that

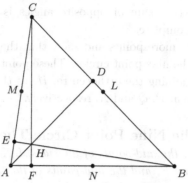

Fig. 3.12   Important points

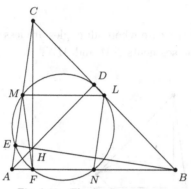

Fig. 3.13   The Six Point Circle.

F lies on this circle. Since we have chosen F arbitrarily as one of the feet of the altitudes, we shall obtain that all of the feet of the altitudes lie on this circle, and the proof will be complete.

Since $\angle CDA$ and $\angle CFA$ are right angles, we have that AC is the diameter of the circle $AFDC$. Since M is the mid-point of AC, we have that M is the centre of the circle $AFDC$. Thus, we have

$$MC = MD = MF = MA \ .$$

This gives that triangle $\triangle MFA$ is an isosceles triangle, and thus, $\angle MAF = \angle MFA$.

We note that LN is parallel to CA. Thus, we have

$$\angle LNB = \angle MAF = \angle MFA \ .$$

Finally, we note that ML is parallel to FN, and this shows that quadrilateral $MLNF$ is a symmetric trapezoid. Since symmetric trapezoids are cyclic quadrilaterals (the sum of opposite angles is equal to two right angles), the proof is complete.

We now look at three more points and show that they also lie on the six point circle (to give the nine point circle). These points are the mid-points of the line segments joining the orthocentre, H, to the vertices A, B and C. We shall name them P, Q and R, respectively.

Theorem 3.12. The Nine Point Circle Theorem

For triangle $\triangle ABC$, the mid-points of the sides, L, M and N, the feet of the altitudes D, E and F, and the mid-points of the line segments joining the orthocentre to the vertices, P, Q and R, all lie on a circle.

Proof.

We give the proof in the case where all angles are less than $\pi/2$.
To help, draw the line segments RM and MN.

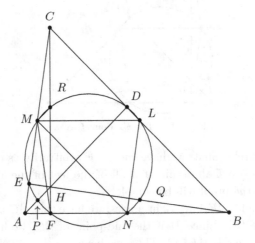

Fig. 3.14   The Nine Point Circle

Since $HR = RC$ and $AM = MC$, we have that RM is parallel to AH, and therefore, that $\angle RML = \angle DAB$.

Since MN is parallel to CB, since LM is parallel to BA, we have that $\angle LMN = \angle MNA = \angle DBA$.

This implies that $\angle RMN$ is a right angle. Since $\angle RFN$ is also a right angle, it follows that $RMFN$ is a cyclic quadrilateral. Since R was chosen arbitrarily amongst P, Q and R, the proof is now complete.

Problem 3.10. *The proof of Theorem 3.11 needs to be modified in the case when* $\angle CAB \geq \pi/2$. *Give a proof for the cases:*

(1) $\angle CAB > \pi/2$;
(2) $\angle CAB = \pi/2$.

Problem 3.11. *The proof of Theorem 3.12 needs to be modified in the case when* $\angle CAB \geq \pi/2$.

(1) *Give a proof for the case:* $\angle CAB > \pi/2$.
(2) *Show that no proof is needed in the case* $\angle CAB = \pi/2$.

3.6 The Euler Line

In this section, we look at the relationship between the orthocentre, the circumcentre, the centroid and the centre of the nine point circle. Much of what is known is due to Euler.

Theorem 3.13.
(a) *The centroid* G, *the circumcentre* O *and the orthocentre* H *are collinear;*
(b) $HG = 2GO$.

Definition 3.6. The line OGH is known as the **Euler line**.

Proof.
To help, we draw the diameter AOX and the line segments XB, XC and ON.

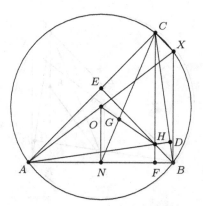

Fig. 3.15   The Euler Line.

We do not assume that OGH is a straight line segment.

First we show that quadrilateral $CXBH$ is a parallelogram.

Since AX is a diameter, we have that $\angle XCA$ is a right angle. Since BE is an altitude, we have that BE is perpendicular to AC. Thus, BH is parallel to CX.

Since AX is a diameter, we have that $\angle XBA$ is a right angle. Since CF is an altitude, we have that CF is perpendicular to AB. Thus, HC is parallel to BX, and we have the parallelogram $CXBH$.

Next we observe that O is the mid-point of AX and that N is the mid-point of AB. From this it follows that triangles $\triangle AON$ and $\triangle AXB$ are similar, and so that $2ON = XB$. But $XB = CH$; therefore, $2ON = CH$.

Since ON is parallel to CF, it follows that $\angle ONG = \angle HCG$. Since G is the centroid, we have that $2NG = GC$. Since we have shown that $2ON = CH$, it follows that triangles $\triangle ONG$ and $\triangle CHG$ are similar.

We now have that $\angle NOG = \angle CHG$, from which it follows that OGH is a straight line segment.

Finally, from the similar triangles, and the known ratio of the corresponding sides, we have that $2OG = GH$. The proof is now complete.

Theorem 3.14. *The centre of the nine point circle is the mid-point of the line joining the circumcentre and the orthocentre.*

Proof.

We add a point and a line segment to the diagram used for the previous theorem: let R be the mid-point of the line segment CH (and thus, lies on the nine point circle), and draw the line segment RN. Suppose that RN and OH intersect at V.

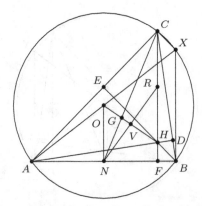

Fig. 3.16   The centre of the Nine Point Circle and the Euler Line

Since $\angle RFN$ is a right angle, it follows that RN is a diameter of the nine point circle.

Since R is the mid-point of CH, it follows that $RH = ON$. Since ON is parallel to RH, we now see that triangles $\triangle ONV$ and $\triangle RHV$ are congruent. This gives that $RV = VN$, so that V is the mid-point of RN, a diameter of the nine point circle. Therefore, V is the centre of the nine point circle.

From the congruent triangles, we also have that $OV = VH$. The proof is now complete.

3.7   The Miquel Point

The last section in this chapter is a little different. All previous concurrence results have involved lines. Here we give a result about the concurrence of three circles. The theorem is due to a nineteenth century French mathematician, A. Miquel.

Theorem 3.15.   Miquel's Theorem.
In triangle $\triangle ABC$, let D, E and F be any points on the sides of the triangle BC, CA and AB, respectively.
Then the three circles, AEF, BFD and CDE are concurrent.

Definition 3.7. The point of intersection of the three circles in Miquel's Theorem is called the **Miquel Point**.

Proof.
We shall consider the circles AEF and BFD, and suppose that they intersect at M (and F). To help, draw the line segments MD, ME and MF.

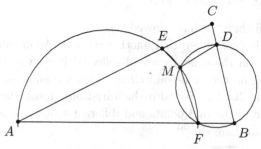

Fig. 3.17   Miquel's Theorem.

Consider angle $\angle EMD$. We have that

$$\angle EMD = 2\pi - \left(\angle EMF + \angle DMF\right) .$$

Since both the quadrilaterals $AEMF$ and $BDMF$ are cyclic, we have that $\angle EMF = \pi - \angle EAF$ and that $\angle DMF = \pi - \angle DBF$. This gives that

$$\angle EMD = \angle EAF + \angle DBF = \angle CAB + \angle CBA .$$

Since the angles of $\triangle ABC$ have sum π, it follows that

$$\angle EMD + \angle ECD = \pi ,$$

and this shows that quadrilateral $CEMD$ is cyclic. The proof is now complete.

Problem 3.12. *Show that the Euler line is perpendicular to one of the sides of a triangle, then the triangle must be isosceles.*

Problem 3.13. *In acute-angled $\triangle ABC$, the circumradius, R, is 8; the distance between the centroid, G, and the centre of the nine-point circle, V, is 1; and $CH = 15$ (H is the orthocentre).*
Show that there is no such $\triangle ABC$.

3.8   Joshua Klehr's Theorem and its Extensions

In May 1999, Steve Sigur, a high school teacher in Atlanta, Georgia, posted on The Math Forum[1], a notice stating that one of his students (Josh Klehr) noticed that

> Given a triangle with mid-point of each side. Through each mid-point, draw a line whose slope is the reciprocal to the slope of the side through that mid-point. These lines concur.

Sigur then stated that "we have proved this".
There was a further statement that another student (Adam Bliss) followed up with a result on the concurrency of reflected lines, with the point of concurrency lying on the nine-point circle. This was subsequently proved by Louis Talman (3.8.1). See also the variations, using the feet of the altitudes in place of the mid-points and different reflections in the recent paper by Floor van Lamoen [2].

---

[1]See http://mathforum.com/epigone/geom.college/grimpcherflend

3.8.1   *First extension*

Here, we are interested in a generalization of Klehr's result.

At the mid-point of each side of a triangle, we construct the line such that the product of the slope of this line and the slope of the side of the triangle is a fixed constant. To make this clear, the newly created lines have slopes of the fixed constant times the reciprocal of the slopes of the sides of the triangle with respect to a given line (parallel to the x–axis used in the Cartesian system). We show that the three lines obtained are always concurrent.

Further, the locus of the points of concurrency is a rectangular hyperbola. This hyperbola intersects the side of the triangles at the mid-points of the sides, and each side at another point. These three other points, when considered with the vertices of the triangle opposite to the point, form a Ceva configuration. Remarkably, the point of concurrency of these Cevians lies on the circumcircle of the original triangle.

Since we are dealing with products of slopes, we have restricted ourselves to a Cartesian proof.

Suppose that we have a triangle with vertices $(0,0)$, $(2a, 2b)$ and $(2c, 2d)$.

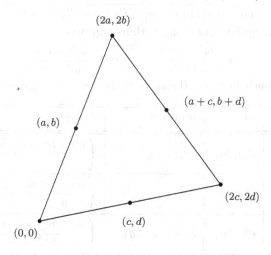

To ensure that the triangle is not degenerate, we insist that $ad - bc \neq 0$. For ease of proof, we also take $0 \neq a \neq c \neq 0$ and $0 \neq b \neq d \neq 0$ to avoid division by zero. However, by continuity, the results obtained here can readily be extended to include these cases.

At the mid-point of each side, we find the equations of the new lines:

Mid-point	Slope	Equation
(a, b)	$\dfrac{\lambda a}{b}$	$y = \dfrac{\lambda a}{b}x + \dfrac{b^2 - \lambda a^2}{b}$
(c, d)	$\dfrac{\lambda c}{d}$	$y = \dfrac{\lambda c}{d}x + \dfrac{d^2 - \lambda c^2}{d}$
$(a + c, b + d)$	$\dfrac{\lambda(c - a)}{d - b}$	$y = \dfrac{\lambda(c - a)}{d - b}x + \dfrac{(a^2 - c^2)\lambda + (d^2 - b^2)}{d - b}$

With the aid of a computer algebra program, we find that the first two lines meet at

$$\left(\frac{\lambda(a^2 d - bc^2) + bd(d - b)}{\lambda(ad - bc)} , \frac{\lambda ac(a - c) + (ad^2 - b^2 c)}{(ad - bc)} \right) ,$$

which it is easy to verify lies on the third line.

By eliminating λ from the equations

$$x = \frac{\lambda(a^2 d - bc^2) + bd(d - b)}{\lambda(ad - bc)} , \qquad y = \frac{\lambda ac(a - c) + (ad^2 - b^2 c)}{(ad - bc)} ,$$

we find that the locus of the points of concurrency is

$$y = \frac{abcd(a - c)(d - b)}{(ad - bc)(x(ad - bc) + (bc^2 - a^2 d))} + \frac{ad^2 - b^2 c}{ad - bc} .$$

This is a rectangular hyperbola, with asymptotes

$$x = \frac{bc^2 - a^2 d}{ad - bc} , \qquad y = \frac{ad^2 - b^2 c}{ad - bc} .$$

Now, this hyperbola meets the sides of the given triangle, as follows:

from	to	mid-point	new-point
$(0, 0)$	$(2a, 2b)$	(a, b)	$\left(\dfrac{ad + bc}{b} , \dfrac{ad + bc}{a} \right)$
$(0, 0)$	$(2c, 2d)$	(c, d)	$\left(\dfrac{ad + bc}{d} , \dfrac{ad + bc}{c} \right)$
$(2a, 2b)$	$(2c, 2d)$	$(a + c, b + d)$	$\left(\dfrac{ad - bc}{d - b} , \dfrac{ad - bc}{a - c} \right)$

The three lines joining the three points (new-point, in each case) to the vertices opposite are concurrent! (Again, easily shown by computer algebra.) The point of concurrency is

$$\left(2(a - c)\left(\frac{ad + bc}{ad - bc} \right) , 2(d - b)\left(\frac{ad + bc}{ad - bc} \right) \right) .$$

It is easy to check that this point is not on the hyperbola. However, it is also easy to check that this point lies on the circumcircle of the original triangle. (Compare this result with the now known result that the point on the hyperbola corresponding to $\lambda = 1$ lies on the nine-point circle. See [1].) In the diagrams that follow on the next four pages, we illustrate the original triangle ABC, the rectangular hyperbola $YWLPXQVZ$ (where $\lambda < 0$) and $MSOUN$ (where $\lambda > 0$), the asymptotes (dotted lines), the circumcircle and the nine-point circle, and the first remarkable point K.

Figure A shows various lines through the mid-points of the sides being concurrent on the hyperbola.

Figure B shows the lines concurrent through the first remarkable point K.

Figure C shows the lines concurrent through the second remarkable point J, where we join points with parameters λ and $-\lambda$.

Figure D shows the parallel lines (or lines concurrent at infinity), where we join points with parameters λ and $-\frac{1}{\lambda}$.

References

[1] Louis A. Talman, *A Remarkable Concurrence*, WEB document, http://clem.mscd.edu/~talmanl, 1999.

[2] Floor van Lamoen, *Morley related triangles on the nine-point circle*, Amer. Math. Monthly **107** (2000), 941–945.

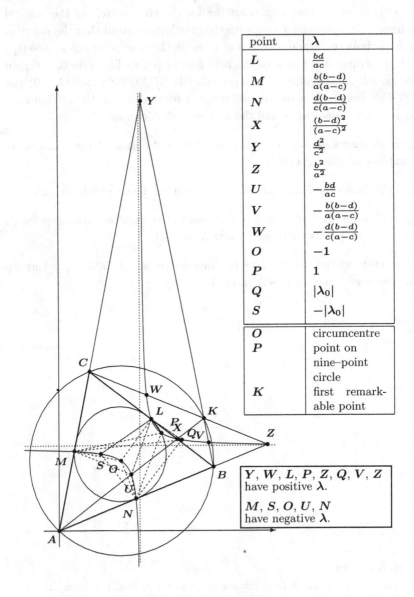

point	λ
L	$\frac{bd}{ac}$
M	$\frac{b(b-d)}{a(a-c)}$
N	$\frac{d(b-d)}{c(a-c)}$
X	$\frac{(b-d)^2}{(a-c)^2}$
Y	$\frac{d^2}{c^2}$
Z	$\frac{b^2}{a^2}$
U	$-\frac{bd}{ac}$
V	$-\frac{b(b-d)}{a(a-c)}$
W	$-\frac{d(b-d)}{c(a-c)}$
O	-1
P	1
Q	$|\lambda_0|$
S	$-|\lambda_0|$

O	circumcentre
P	point on nine–point circle
K	first remarkable point

Y, W, L, P, Z, Q, V, Z have positive λ.

M, S, O, U, N have negative λ.

Fig. A.   Concurrences with variable values of λ.

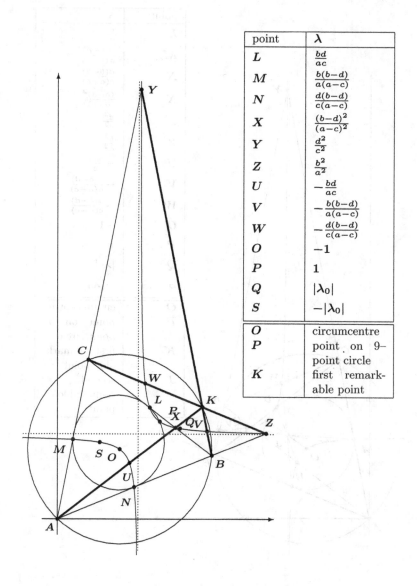

point	λ
L	$\frac{bd}{ac}$
M	$\frac{b(b-d)}{a(a-c)}$
N	$\frac{d(b-d)}{c(a-c)}$
X	$\frac{(b-d)^2}{(a-c)^2}$
Y	$\frac{d^2}{c^2}$
Z	$\frac{b^2}{a^2}$
U	$-\frac{bd}{ac}$
V	$-\frac{b(b-d)}{a(a-c)}$
W	$-\frac{d(b-d)}{c(a-c)}$
O	-1
P	1
Q	$|\lambda_0|$
S	$-|\lambda_0|$
O	circumcentre
P	point on 9–point circle
K	first remarkable point

Fig. B.   Concurrences through first remarkable point K.

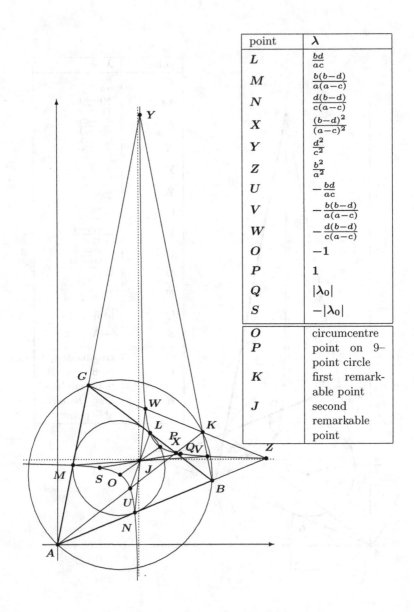

point	λ
L	$\frac{bd}{ac}$
M	$\frac{b(b-d)}{a(a-c)}$
N	$\frac{d(b-d)}{c(a-c)}$
X	$\frac{(b-d)^2}{(a-c)^2}$
Y	$\frac{d^2}{c^2}$
Z	$\frac{b^2}{a^2}$
U	$-\frac{bd}{ac}$
V	$-\frac{b(b-d)}{a(a-c)}$
W	$-\frac{d(b-d)}{c(a-c)}$
O	-1
P	1
Q	$|\lambda_0|$
S	$-|\lambda_0|$

O	circumcentre
P	point on 9-point circle
K	first remarkable point
J	second remarkable point

Fig. C.   Concurrences through second remarkable point J.
Points with parameters λ and $-\lambda$.

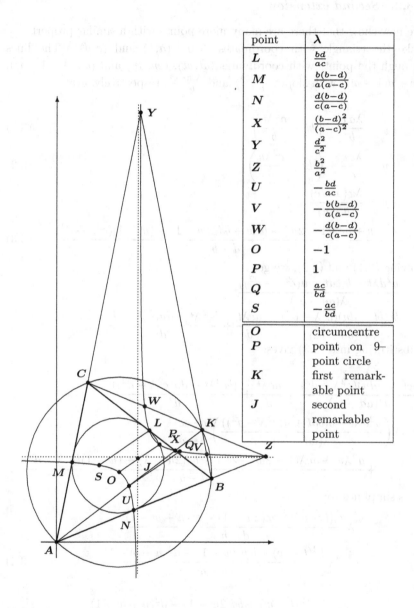

point	λ
L	$\frac{bd}{ac}$
M	$\frac{b(b-d)}{a(a-c)}$
N	$\frac{d(b-d)}{c(a-c)}$
X	$\frac{(b-d)^2}{(a-c)^2}$
Y	$\frac{d^2}{c^2}$
Z	$\frac{b^2}{a^2}$
U	$-\frac{bd}{ac}$
V	$-\frac{b(b-d)}{a(a-c)}$
W	$\frac{d(b-d)}{c(a-c)}$
O	-1
P	1
Q	$\frac{ac}{bd}$
S	$-\frac{ac}{bd}$

O	circumcentre
P	point on 9-point circle
K	first remarkable point
J	second remarkable point

Fig. D.   Parallel lines — concurrences at infinity.
Points with parameters λ and $-\frac{1}{\lambda}$.

3.8.2 *Second extension*

We now show that there are many more points with a similar property.
Take the triangle with coordinates $(0,0)$, (a,b) and (c,d). The lines
through the points with coordinates (at, bt), (cu, du) and $(aw + c(1 - w),$
$bw + d(1 - w))$ with slopes $\frac{\lambda a}{b}$, $\frac{\lambda c}{d}$ and $\frac{\lambda(a-c)}{b-d}$, respectively, are:

$$y = \frac{\lambda a x}{b} + \left(bt - \frac{a^2 \lambda t}{b} \right) \tag{3.1}$$

$$y = \frac{\lambda c x}{d} + \left(du - \frac{c^2 \lambda u}{d} \right) \tag{3.2}$$

$$y = \frac{\lambda x (a - c)}{(b - d)}$$
$$+ \frac{a^2 \lambda w + ac\lambda(1 - 2w) - b^2 w + bd(2w - 1) + (w - 1)(c^2 \lambda - d^2)}{d - b} . \tag{3.3}$$

Solving (3.1) and (3.2), we get
$$x = \frac{a^2 d\lambda t - b(bdt + u(c^2\lambda - d^2))}{\lambda(ad - bc)}$$
$$y = \frac{bc(bt - du)(c^2\lambda - d^2)}{(d^2(ad - bc))} + \frac{ac\lambda t}{d} + \frac{bc^2\lambda t - du(c^2\lambda - d^2)}{d^2} .$$
Substituting into (3.3) gives:

$$\frac{bc(bt - du)(c^2\lambda - d^2)}{d^2(ad - bc)} + \frac{ac\lambda t}{d} + \frac{(bc^2\lambda t - du(c^2\lambda - d^2))}{d^2}$$
$$= \lambda \left(\frac{a^2 d\lambda t - b(bdt + u(c^2\lambda - d^2))}{\lambda(ad - bc)} \right) \frac{a - c}{b - d}$$
$$+ \frac{a^2\lambda w + ac\lambda(1 - 2w) - b^2 w + bd(2w - 1) + (w - 1)(c^2\lambda - d^2)}{d - b} .$$

This simplifies to:
$$\lambda \frac{a^2(t - w) + ac(2w - 1) - c^2(u + w - 1)}{d - b}$$
$$+ \frac{b^2(t - w) + bd(2w - 1) - d^2(u + w - 1)}{b - d} = 0, \tag{3.4}$$

or

$$\lambda = \frac{b^2(t - w) + bd(2w - 1) - d^2(u + w - 1)}{a^2(t - w) + ac(2w - 1) - c^2(u + w - 1)}. \tag{3.5}$$

If $t = u = w = \frac{1}{2}$, then left side of (3.4) is zero. This means that the lines
always concur.

If we make the same substitution into (3.4), we get:

$$\lambda = \frac{b^2(\frac{1}{2} - \frac{1}{2}) + bd(2(\frac{1}{2}) - 1) - d^2(\frac{1}{2} + \frac{1}{2} - 1)}{a^2(\frac{1}{2} - \frac{1}{2}) + ac(2(\frac{1}{2}) - 1) - c^2(\frac{1}{2} + \frac{1}{2} - 1)} = \frac{0}{0}.$$

There are other values of u, v and w that result in the same result. Note that we will have two linear equations in the three unknowns t, u and w. Thus, unless a, b, c and d are "peculiar", we have infinitely many such triples.

We call the set of all values of u, v and w that make the left side of (3.5) zero, the **Concurrence Set** of the triangle. This is given by

$$\frac{T}{W} = 1 - \frac{2cd}{ad + bc},$$
$$\frac{U}{W} = \frac{2ab}{ad + bc} - 1,$$

where $t = \frac{1}{2} - T$, $u = \frac{1}{2} - U$ and $w = \frac{1}{2} - W$. Thus, there are infinitely many points in this set.

For points not in the concurrence set, we have that there is one unique value of λ such that the three lines concur.

The value is
$$\frac{b^2(t - w) + bd(2w - 1) - d^2(u + w - 1)}{a^2(t - w) + ac(2w - 1) - c^2(u + w - 1)}.$$

Replace t by $\frac{1}{2} - T$, etc.; the value is
$$\frac{b^2(T - W) + 2bdW - d^2(U + W)}{a^2(T - W) + 2acW - c^2(U + W)}.$$

The point of concurrence is
$$x = \frac{N_x}{D_x}, \text{ where}$$

$$N_x = (TU(2ad(b - d) + 2bc(b - d))$$
$$+ TW(2ad(2b - d) - 2bcd) - bT(ad + bc)$$
$$- 2bUW(ad + c(b - 2d))$$
$$+ U(ad^2 + bcd) + W(d - b)(ad - bc)),$$

and
$$D_x = (2b^2T - 2d^2U - 2W(b^2 - 2bd + d^2));$$

and $y = \dfrac{N_y}{D_y}$, where

$$N_y = (2TU(a^2d + ac(b - d) - b^2c^2)$$
$$+2cTW(a(2b - d) - b^2c) - aT(ad + bc)$$
$$-2aUW(ad + c(b - 2d))$$
$$+cU(ad + b^2c) + W(a^2d - ac(b + d) + b^2c^2))$$

and

$$D_y = (2a^2T - 2c^2U - 2W(a^2 - 2ac + c^2)).$$

For fixed T and U, or any other pair, the locus of this may reduce to a single point or to a rectangular hyperbola, or even a straight line.

Chapter 4

Circumcircle, Inradius and Semiperimeter Formulae

4.1 Semi-perimeter Formulae

The perimeter of a plane figure is the length of its boundary. The semi-perimeter is, of course, one half of the perimeter. The standard notation for a triangle is to denote the the semi-perimeter by the symbol s, so that

$$s = \frac{a+b+c}{2} . \tag{4.1}$$

Many results about the semi-perimeter also involve the radius of the inscribed circle, usually called the in-radius, and usually denoted by r. The centre, I, of the inscribed circle (or incircle) is obtained at the intersection of the bisectors of the angles of the triangle.

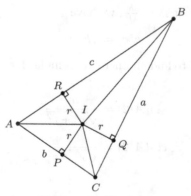

Fig. 4.1 The Incentre.

Let P, Q and R be the feet of the perpendiculars from I to BC, CA and AB, respectively. Then $IP = IQ = IR = r$.

Note that $AQ = AR$, $BR = BP$ and $CP = CQ$, so that $s = AR+BP+CQ$, etc. Note that this gives $AR = s - a$, etc.

The Ravi Substitution We shall define the **Ravi Substitution** [1] using $AR = x$, $BP = y$ and $CQ = z$. Thus,

$$x = s - a \quad y = s - b \quad z = s - c \ ,$$

so that $x + y + z = s$, $x + y = c$, etc.

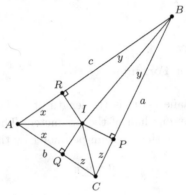

Fig. 4.2   The Ravi Substitution

Note that the area of quadrilateral $ARIQ$ is equal to rx. Thus, we get

$$\Delta = r(x + y + z) = rs \ .$$

Since we also know that $\Delta = \frac{abc}{4R}$, we have

$$abc = 4Rrs \ .$$

We shall now proceed to find another two formulae for the area of a triangle. First we note that

$$\sin(A/2) = \frac{r}{\sqrt{x^2 + r^2}} \ ,$$

$$\cos(A/2) = \frac{x}{\sqrt{x^2 + r^2}} \ .$$

Thus, we have that

$$\sin(A) = 2\sin(A/2)\cos(A/2) = \frac{2xr}{x^2 + r^2} \ .$$

---

[1] A favourite of the Canadian Mathematician, Ravi Vakil

Therefore,

$$\Delta \; = \; \frac{1}{2}bc\sin(A) \; = \; \frac{bcxr}{x^2 + r^2} \; .$$

Substituting for r from $\Delta = rs$, we get

$$\Delta = \frac{bcx\Delta/s}{x^2 + (\Delta/s)^2} \; ,$$

or $\qquad\qquad x^2 + \dfrac{\Delta^2}{s^2} = \dfrac{xbc}{s} \; ,$

so that $\qquad\qquad \dfrac{\Delta^2}{s} = x(bc - xs)$

$$= x\left((x+z)(y+x) - x(x+y+z)\right)$$

$$= xyz \; .$$

This gives rise to the well known formulae

$$\Delta \; = \; \sqrt{(x+y+z)xyz} \; = \; \sqrt{s(s-a)(s-b)(s-c)} \; .$$

We also get formulae for the inradius

$$r \; = \; \sqrt{\frac{xyz}{(x+y+z)}} \; = \; \sqrt{\frac{(s-a)(s-b)(s-c)}{s}} \; .$$

4.2 Half Angle Formulae

We recall two formulae from above:

$$\sin(A/2) \; = \; \frac{r}{\sqrt{x^2 + r^2}} \; , \qquad \cos(A/2) \; = \; \frac{x}{\sqrt{x^2 + r^2}} \; .$$

Since we know that

$$r \; = \; \sqrt{\frac{xyz}{(x+y+z)}} \; ,$$

we can calculate that

$$x^2 + r^2 \; = \; \frac{x(x+y)(x+z)}{(x+y+z)} \; = \; \frac{xbc}{s} \; .$$

Using these, we get

$$\sin(A/2) = \sqrt{\frac{xyz}{(x+y+z)}} \cdot \sqrt{\frac{(x+y+z)}{x(x+y)(x+z)}}$$

$$= \sqrt{\frac{yz}{(x+y)(x+z)}}$$

$$= \sqrt{\frac{(s-b)(s-c)}{bc}} \quad ,$$

$$\cos(A/2) = x\sqrt{\frac{(x+y+z)}{x(x+y)(x+z)}}$$

$$= \sqrt{\frac{x(x+y+z)}{(x+y)(x+z)}}$$

$$= \sqrt{\frac{s(s-a)}{bc}} \quad ,$$

$$\tan(A/2) = \sqrt{\frac{yz}{x(x+y+z)}}$$

$$= \sqrt{\frac{(s-b)(s-c)}{s(s-a)}} \quad .$$

There are, of course, similar formulae for the other angles.

Problem 4.1. *Prove that* $s = 4R\cos(A/2)\cos(B/2)\cos(C/2)$. (Solution on page 188.)

Problem 4.2. *Prove that* $\Delta = s^2\tan(A/2)\tan(B/2)\tan(C/2)$. (Solution on page 188.)

4.3   In-radius Formulae

We already have obtained some formulae that involve the in-radius r:

$$r = \frac{\Delta}{s}$$

$$= \frac{abc}{4Rs}$$

$$= \sqrt{\frac{xyz}{x+y+z}} \ .$$

However, there are other formulae, easily seen from the diagram:

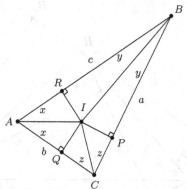

Fig. 4.3   In-radius Formulae.

$$r = (s-a)\tan(A/2) \ = \ x\tan(A/2)$$
$$= (s-b)\tan(B/2) \ = \ y\tan(B/2)$$
$$= (s-c)\tan(C/2) \ = \ z\tan(C/2) \ .$$

Problem 4.3. *Prove that* $r = s\tan(A/2)\tan(B/2)\tan(C/2)$. (Solution on page 188.)

Problem 4.4. *Prove that* $r = 4R\sin(A/2)\sin(B/2)\sin(C/2)$. (Solution on page 189.)

Problem 4.5. *Prove that* $\Delta \ = \ r^2\cot(A/2)\cot(B/2)\cot(C/2)$. (Solution on page 189.)

4.4 Escribed Circles

The inscribed circle touches all three line segments that form the sides of a triangle **internally**. Escribed circles touch two of the line segments that form the sides of a triangle, **extended**. Their centres are at the intersection of two **external** bisectors of the angles of the triangle, with the internal bisector of the third. (There are, of course, three escribed circles.) The diagram below shows the centre (I_A) of the one of them opposite angle A, with radius r_A.

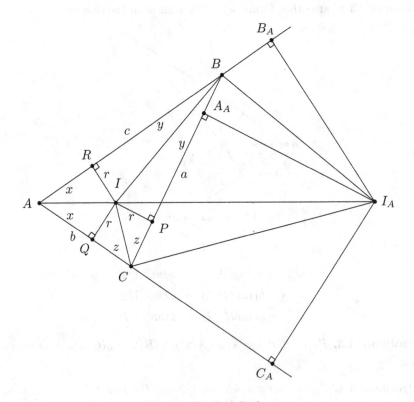

Fig. 4.4   Escribed Circles

Problem 4.6. *Prove that* $BA_A = z$.

Problem 4.7. *Prove that:* (Solution on page 190.)

$$r_A = s \tan(A/2) = (x + y + z) \tan(A/2)$$
$$= (s - c) \cot(B/2) = z \cot(B/2)$$
$$= (s - b) \cot(C/2) = y \cot(C/2)$$
$$= \frac{a \cos(B/2) \cos(C/2)}{\cos(A/2)}$$
$$= \Delta/(s - a) = \Delta/x \ .$$

There are, of course, similar formulae for r_B and r_C.

Problem 4.8. *Suppose that R is the circumradius of triangle ABC and r_A, r_B and r_C are the radii of the escribed circles. Show that*

(1) $r_A r_B + r_B r_C + r_C r_A = s^2$,
(2) $\Delta ABC = \frac{r_A r_B r_C}{s}$.

If the distances between the centres of the escribed circles are α, β and γ, and $\sigma = \frac{\alpha + \beta + \gamma}{2}$ show that

$$8R = \frac{\alpha \beta \gamma}{\sqrt{\sigma(\sigma - \alpha)(\sigma - \beta)(\sigma - \gamma)}} \ .$$

(Solution on page 191.)

Chapter 5

Conic Sections

5.1 Review

We assume a certain familiarity with conic sections. For example, we assume the descriptive approach of cutting a (double) cone with a plane:

- If the plane is parallel to the slant of the cone (and therefore cuts only one part of the double cone), the section is a **parabola**.
- If the plane cuts both parts of the double cone, the section is a **hyperbola**.

  In particular, if the plane is parallel to the axis of the cone (and therefore cuts both parts of the double cone), the section is a **hyperbola**.
- If the plane cuts only one part of the double cone, the section is an **ellipse**.

  In particular, if the plane is perpendicular to the axis of the cone, the section is a **circle**.

5.2 Basic Definitions

We shall define conic sections with a synthetic approach. We shall deduce their Cartesian equations and indicate some properties.

Definition 5.1. A **conic section** is the locus of a point P, which satisfies the following property:
the distance of P from a fixed point F is equal to a fixed non-negative number e times the distance of P from a fixed line l.
The fixed point F is called the **focus**.

85

The fixed line l is called the **directrix**.
The fixed number e is called the **eccentricity**.

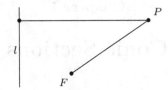

Fig. 5.1   Focus and Directrix.

Depending on the size of the number e, we get different conic sections:

- If $e < 1$, we have an **ellipse**.
  In particular, if $e = 0$, we have a **circle**.
- If $e = 1$, we have a **parabola**.
- If $e > 1$, we have a **hyperbola**.

5.3   The Parabola

Here, the eccentricity, e, has value 1.

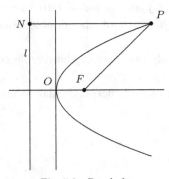

Fig. 5.2   Parabola.

Suppose that P has coordinates (x, y), that l has equation $x = -p$, and that F has coordinates $(p, 0)$,
Then N (the foot of the perpendicular from P to l) has coordinates $(-p, y)$.
Thus, $NP = x + p$.
Using the distance formula, we have that

$$PF^2 = (x - p)^2 + (y - 0)^2 .$$

Since $NP = FP$, we now have

$$(x+p)^2 = (x-p)^2 + y^2 ,$$

which leads to the standard equation for a parabola:

$$y^2 = 4px .$$

Problem 5.1. *A parabola has focus F and directrix l.*
The chord PQ passes through the focus F. L and M lie on l and satisfy
LP $\perp$ l and MQ $\perp$ l. The point N is such that LN$\|$PF and NF$\|$LP.
Suppose that PN intersects LM at R.

(1) Prove that $\angle LFM$ is a right angle;
(2) Prove that R is the mid-point of LM.

5.4　The Ellipse

Here, the eccentricity, e, has value less than 1. For a circle, the value of
e is zero. This means that the directrix is a "line at infinity", and the
definition, in a sense, breaks down. We shall therefore develop ellipses only
in the cases $0 < e < 1$.

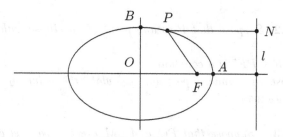

Fig. 5.3　Ellipse.

Suppose that P has coordinates (x, y), that F has coordinates $(ae, 0)$ and
that the line l has equation $x = a/e$. (We take, as usual, $a > 0$.)
Then $PN = a/e - x$ and $PF^2 = (ae - x)^2 + y^2$. This reduces to:

$$\frac{x^2}{a^2} + \frac{y^2}{a^2(1 - e^2)} = 1 ,$$

which is the standard equation for an ellipse. We make the substitution
$b = a\sqrt{1 - e^2}$, to get the normal form:

$$\frac{x^2}{a^2} + \frac{y^2}{b^2} = 1 .$$

The numbers a and b represent half the lengths of the major and minor axes, and are referred to as the **semi-major** and **semi-minor** axes. On the diagram, they are given by OA and OB.

Because the equation obtained for an ellipse is symmetric both in x and in y, we can see that it can be generated from another point as focus and another line as directrix. The point and the line are the mirror images in the y–axis of the focus and the directrix. These are sometimes called the anti-focus and the anti-directrix, and denoted by F' and l'. Usually, F and F' are called the foci of the ellipse, and l and l', the directrices of the ellipse.

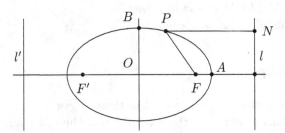

Fig. 5.4   Ellipse with foci and directrices.

Problem 5.2. *Suppose that P is any point of an ellipse with foci F and F'.*
Show that $PF + PF'$ is a constant.
For an ellipse in standard position, calculate the value of $PF + PF'$.
(Solution on page 195.)

Problem 5.3.   *Suppose that PQ and LM are two parallel chords of an ellipse.*
Suppose that R and N are the mid-points of PQ and LM, respectively.
Show that the line RN passes through the centre of the ellipse. (Solution on page 196.)

5.5   The Hyperbola

Here, the eccentricity, e, has value greater than 1.
Suppose that P has coordinates (x, y), that F has coordinates $(ae, 0)$ and that the line l has equation $x = a/e$. (We take, as usual, $a > 0$.)

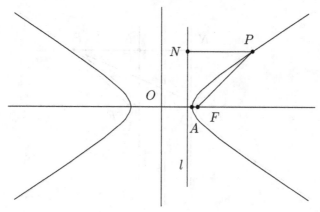

Fig. 5.5 Hyperbola.

Then $PN = x - a/e$ and $PF^2 = (ae - x)^2 + y^2$. This reduces to:

$$\frac{x^2}{a^2} - \frac{y^2}{a^2(e^2 - 1)} = 1 \,,$$

which is the standard equation for a hyperbola. We make the substitution $b = a\sqrt{e^2 - 1}$, to get the normal form:

$$\frac{x^2}{a^2} - \frac{y^2}{b^2} = 1 \,.$$

The number a represents half the length of the of the distance between the two "noses" of the hyperbola. This is sometimes called the **semi-major** axis. On the diagram, it is given by OA.

Unlike an ellipse, there is no obvious interpretation for a **semi-minor** axis. However, if we were to draw, on the same diagram, the **conjugate** hyperbola, given by

$$\frac{x^2}{b^2} - \frac{y^2}{a^2} = 1 \,,$$

we would find that it had semi-major axis of length b.

Because the equation obtained for a hyperbola is symmetric both in x and in y, we can see that it can be generated from another point as focus and another line as directrix. The point and the line are the mirror images in the y–axis of the focus and the directrix. These are sometimes called the anti-focus and the anti-directrix, and denoted by F' and l'. Usually, F and

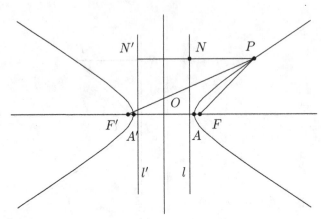

Fig. 5.6   Hyperbola with foci and directrices.

F' are called the foci of the hyperbola, and l and l', the directrices of the hyperbola.

Problem 5.4. *Suppose that F and F' are the foci of a hyperbola and that P is any point on the hyperbola.*
Calculate the set of values taken by $|PF| - |PF'|$. ($|PF|$ means the length of the line segment PF). (Solution on page 197.)

Associated with the hyperbola in standard form $x^2/a^2 - y^2/b^2 = 1$, there is a curve given by the equation $x^2/a^2 - y^2/b^2 = 0$. This curve is in fact a pair of lines passing through the origin O and with slopes $\pm b/a$. These lines are **asymptotic** to the hyperbola as $|x| \to \infty$ and as such as called the **asymptotes of the hyperbola**.
The asymptotes to the hyperbola $x^2/a^2 - y^2/b^2 = 1$ are also the asymptotes to the conjugate hyperbola $x^2/b^2 - y^2/a^2 = 1$.

In the special case $a^2 = b^2$, the asymptotes are at right angles to one another, and the hyperbola is then called a **rectangular** hyperbola. It is usual to draw a rectangular hyperbola with the axes as its asymptotes.

Problem 5.5. *Calculate the eccentricity for the rectangular hyperbola $xy = 4$.* (Solution on page 197.)

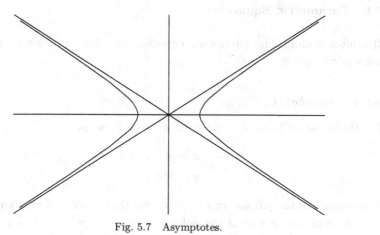

Fig. 5.7   Asymptotes.

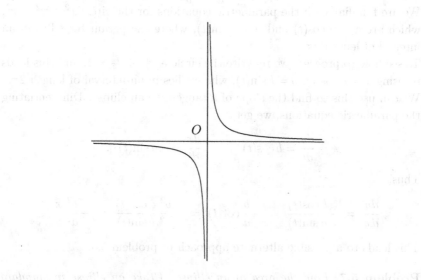

Fig. 5.8   Rectangular hyperbola.

5.6   Parametric Equations

It is often useful to use parametric equations in solving problems involving conic sections.

5.6.1   *Parabolas*

For the standard form of a parabola, $y^2 = 4px$, we use

$$x = pt^2 \ ,$$
$$y = 2pt \ .$$

The domain of the parameter t is $(-\infty, \infty)$; that is, all real numbers. The slope of the tangent at the point $P = (x, y) = (pt^2, 2pt)$ is found from differentiating $y^2 = 4px$, yielding $2yy' = 4p$, or $y' = \frac{2p}{y} = \frac{1}{t}$.

Problem 5.6. *If $P = (x, y) = (pt^2, 2pt)$, prove that the chord on FP cuts the parabola again when $x = \frac{p}{t^2}$, $y = \frac{2p}{t}$.* (Solution on page 198.)

5.6.2   *Ellipses*

The standard form of an ellipse is $\dfrac{x^2}{a^2} + \dfrac{y^2}{b^2} = 1$.

We are familiar with the parametric equations for the circle $x^2 + y^2 = r^2$, which are $x = r\cos(t)$ and $y = r\sin(t)$, where the parameter t lies in an interval of length 2π.

To see how to proceed, we re-write the circle as $\frac{x^2}{r^2} + \frac{y^2}{r^2} = 1$, and this leads to using $x = a\cos(t)$, $y = b\sin(t)$, where t lies in an interval of length 2π. We can use this to find the slope of a tangent to an ellipse. Differentiating the parametric equations, we get

$$\frac{dy}{dt} = b\cos(t) \ , \qquad \frac{dx}{dt} = -a\sin(t) \ .$$

Thus,

$$\frac{dy}{dx} = -\frac{b\cos(t)}{a\sin(t)} = -\frac{b}{a}\cot(t) = -\frac{b^2\,a\cos(t)}{a^2\,b\sin(t)} = -\frac{b^2\,x}{a^2\,y} \ .$$

This leads to a possible alternate approach to problem 5.3.

Problem 5.7. *Find the area of an ellipse. (Take an ellipse in standard form.)* (Solution on page 198.)

5.6.3 *Hyperbolas*

What made the trigonometric functions good for ellipses (and circle) was that fact that $\cos(x)$ and $\sin(x)$ satisfied Pythagoras' Theorem: $a^2 + b^2 = 1$. What we need here are a pair of functions that satisfy a similar result: $a^2 - b^2 = 1$. The answer is given by the hyperbolics functions (they were not so named without a good reason).

The standard form of a hyperbola is $\frac{x^2}{a^2} - \frac{y^2}{b^2} = 1$.

This leads to using $x = a\cosh(t)$, $y = b\sinh(t)$, where t is a real number. Again, we can find the slope of the tangent from

$$\frac{dy}{dx} = \frac{b\cosh(t)}{a\sinh(t)} = \frac{b}{a}\coth(t) = \frac{b^2}{a^2}\frac{a\cosh(t)}{b\sinh(t)} = \frac{b^2}{a^2}\frac{x}{y} .$$

5.7 General Second Degree Equations

All the equations that we have met in the earlier part of this chapter are of second degree in both x and y. It should not be too surprising, then, to learn that every equation of second degree in x and y gives rise to a conic. This can be seen using some standard techniques from Linear Algebra. The general second degree equation can be written in the form:

$$ax^2 + 2hxy + by^2 + 2gx + 2fy + c = 0 .$$

It is written in this form because we can then re-write it in matrix form:

$$(x \ y \ 1) \begin{pmatrix} a & h & g \\ h & b & f \\ g & f & c \end{pmatrix} \begin{pmatrix} x \\ y \\ 1 \end{pmatrix} = 0 ,$$

which we shall abbreviate as

$$\bar{x}^T A \, \bar{x} = 0 .$$

We can now apply the usual techniques to rotate the coordinate system by using matrices of the form

$$U = \begin{pmatrix} \cos(\theta) & \sin(\theta) & 0 \\ -\sin(\theta) & \cos(\theta) & 0 \\ 0 & 0 & 1 \end{pmatrix} .$$

Problem 5.8. *Find the conditions of the numbers a, b, c, f, g and h for the general equation of second degree*

$$ax^2 + 2hxy + by^2 + 2gx + 2fy + c = 0$$

to be a circle. (Solution on page 198.)

Problem 5.9. *For what values of h does the general equation of second degree*

$$ax^2 + 2hxy + by^2 + 2gx + 2fy + c = 0$$

represent a pair of lines? (Solution on page 199.)

Problem 5.10. *Suppose that A and B are two distinct fixed points. Find the locus of the point P which has the property that*

$$\angle PAB - \angle PBA$$

is constant.

What relation does the line segment AB have with this locus? (Solution on page 199.)

Chapter 6

Constructions

In this chapter, we discuss two problems that involve construction with straight edge and compasses alone. They are
  • Regular Polygons.   • Real Numbers.
We shall start with some elementary geometric constructions.

6.1 Elementary Constructions

The constructions detailed here are:

(1) Bisection of a line segment.
(2) Bisection of an angle.
(3) Drawing a perpendicular to a line.
(4) Drawing a parallel line.
(5) Trisection of a line segment.

6.1.1 *Bisection of a line segment*

We are given a line segment AB and wish to find the mid-point C.

Construction
Take the compasses with centre A and radius greater than one half of the length of the line segment AB. In practice, a length of about the length of the line segment AB is taken.
Draw an arc of a circle, about the size of a semi-circle, placed on the same side of A as is B, approximately equally on each side of the line segment AB.

95

With the same radius, repeat this with centre B so that the two arcs intersect at two points: C and D.

Draw the line segment CD and let it intersect AB at M.

Then M is the mid-point of AB.

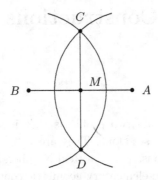

Fig. 6.1   The mid-point of a line segment – 1.

Proof.

Draw the line segments CA, AD, DB and BC.

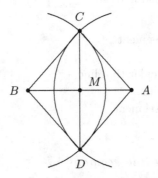

Fig. 6.2   The mid-point of a line segment – 2.

Since $AC = AD = BC = BD$ (all drawn with the same radius), and $CD = CD$, we have $\triangle ACD \equiv \triangle BCD$ (three sides – SSS).

Hence, $\angle ACD = \angle BCD$; that is, $\angle ACM = \angle BCM$.

Since $AC = BC$, $CM = CM$ and $\angle ACM = \angle BCM$, we have that $\triangle ACM \equiv \triangle BCM$ (SAS). Hence, $AM = MB$; that is, M is the mid-point of AB.

6.1.2 *Bisection of an angle*

We start with an angle $\angle BAC$. In the diagrams, we shall take an angle less than a right angle, although the technique is the same for an angle between a right angle and a straight angle. For an angle greater than a straight angle but less than a complete revolution, we work with the "opposite" angle.

The construction begins by drawing an arc, centre A with suitable (reasonable) radius, to intersect BA at P and CA at Q. With the same radius and centres P and Q, draw arcs to intersect at M. (If the two arcs do not intersect, then start again with a larger radius until two intersecting arcs are obtained).

Then the line segment MA bisects the angle $\angle BAC$.

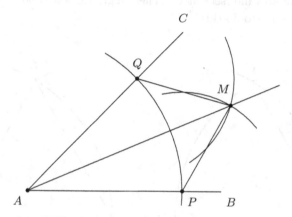

Fig. 6.3 Bisection of an angle.

We can prove that we have the angle bisector by considering triangles $\triangle MPA$ and $\triangle MQA$. Here we have:

$$MA = MA;$$
$$MP = MQ \qquad \text{equal radii in construction;}$$
$$AP = AQ \qquad \text{equal radii in construction.}$$

Thus, triangles $\triangle MPA$ and $\triangle MQA$ are congruent by (SSS), and it therefore follows that

$$\angle MAP = \angle MAQ \,.$$

6.1.3 *Drawing a perpendicular to a line*

There are two different constructions here:

(1) drawing a perpendicular to a line at a point on the line; and
(2) drawing a perpendicular to a line from a point not on the line.

The first is equivalent to the construction of a right angle. Now, a right angle consists of the sum of an angle of 60° with an angle of 30°. And an angle of 30° is half of an angle of 60°, we need only show how to construct an angle of 60°.
Given a line segment AB, with centre A and radius AB, draw an arc of a circle on one side of AB. Repeat this with the same radius and centre B so that the arcs intersect at C. Then angle $\angle CAB = 60°$. Now, can you find other ways to do this?

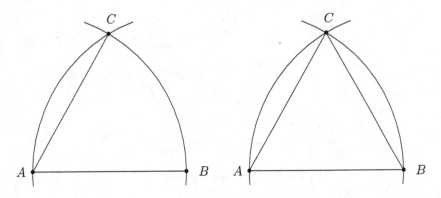

Fig. 6.4   Drawing an angle of 60°.

If we consider the triangle $\triangle ABC$, we see that $AB = BC = CA$: that is, we have an equilateral triangle. Thus, **all** angles are equal to 60°.

We now consider a line AB and a point C not on the line.
We begin by drawing an arc of a circle, centre C to intersect AB at two points P and Q.
With centres P and Q, draw arcs with the same radius to intersect on the other side of AB from C.
Suppose that these arcs intersect at D.
Then CD is perpendicular to AB.

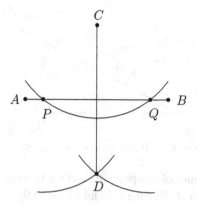

Fig. 6.5 Drawing a perpendicular from a point not on the line.

Problem 6.1. *In the above construction, prove that* $CD \perp AB$. (Solution on page 200.)

6.1.4 *Drawing a parallel line.*

We start with a line AB and a point C not on it. We wish to draw a line through C parallel to AB.

The technique is to apply the previous results. First, we draw a perpendicular line through C to AB. Then we draw, at C, a perpendicular to the new line. This will give the line through C, parallel to AB.

6.1.5 *Trisection of a line segment.*

The construction here is of a different type from that used to bisect a line segment.

We start with the line segment AB to be trisected. At A, we draw a line segment AM at an angle of about half a right angle to AB, in the same direction as is B from A. On this line mark three points as follows:

(1) With centre A and radius about $1/3$ of AB, draw an arc of a circle to cut the line at A_1.
(2) With centre A_1 and the same radius, draw an arc of a circle away from A to cut the line at A_2.
(3) With centre A_2 and the same radius, draw an arc of a circle away from A_1 to cut the line at A_3.

Draw the line segment A_3B.

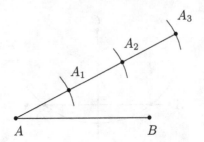

Fig. 6.6   Trisection of a line segment − 1.

Now use the techniques of the previous section to draw line segments A_2C_2 and A_1C_1 parallel to A_3B, with C_2 and C_1 on AB.

The points C_1 and C_2 are the trisection points of AB. This is easily proved by using similar triangles.

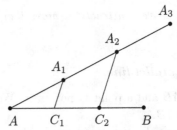

Fig. 6.7   Trisection of a line segment − 2.

This technique is easily extended to allow the division of a line into n equal segments, where n is any positive integer.

Problem 6.2. *Given an angle $\theta > 0$,*

(1) show how to construct an angle of 2θ using straight edge and compasses alone;

(2) show how to construct an angle of $n\theta$ ($n \in \mathbb{N}$) using straight edge and compasses alone.

6.2   Regular Polygons

We ask the question:

Which regular polygons can we construct with straight edge and compasses?

We assume either that we are given one edge, or that we are given a circle into which we wish to inscribe the polygon.

Both assumptions lead to the identical question:

For which integers N, can we construct with straight edge and compasses, the angle of measure $\frac{2\pi}{N}$?

Since we know that we can bisect any angle, we know that we can find certain proportions of a given angle: these proportions are:

$$\frac{m}{2^n} \qquad \text{where } n \in \mathbb{N}, \quad 0 < m < 2^n \ .$$

This leaves the question of what other divisions of an angle are possible. The answer is that we can divide an angle into p equal parts if and only if

$$p = 2^n p_1 p_2 \cdots p_k$$

where each p_j is a distinct **Fermat** prime: that is a prime number of the form

$$2^h + 1 \ .$$

This result was proved by Gauss. The proof of this is beyond the scope of this book.

We shall therefore concentrate on the construction of polygons with a **prime** number of sides. These are then those with

$$3, \ 4, \ 5, \ 17, \ \ldots \text{ sides.}$$

6.2.1 *Three Sides – equilateral triangle*

The construction technique is obtained from what is given above: we need to construct an angle of 60°.

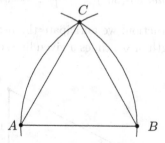

Fig. 6.8   Drawing an angle of 60°.

6.2.2 *Four Sides – square*

Here, we need to be able to construct a right angle. Again, this has been done before. Then we repeat the process.

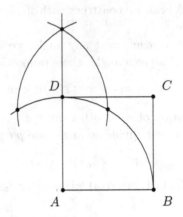

Fig. 6.9   Drawing a square.

6.2.3   *Five Sides – regular pentagon*

The essence here is to construct an angle of 36°. From this we can construct
an angle of 72° and an angle of 54°. To do this, we make use here of the
property that

$$\cos(36°) = \frac{1 + \sqrt{5}}{4} .$$

We should check that this is correct!

Problem 6.3. *Show that* $\cos(36°) = \frac{1+\sqrt{5}}{4}$. (Solution on page 201.)

To complete the construction, we anticipate the next part of this chapter,
and observe that a length of $\sqrt{5}$ times a given line segment is constructible.

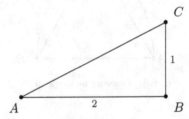

Fig. 6.10   Obtaining a length of $1 + \sqrt{5}$.

If BC is of length 1, if AB is of length 2 and if angle $\angle ABC$ is a right
angle, then Pythagoras' Theorem gives us that AC is of length $\sqrt{5}$.

We can "add" a length of 1 to this with straight edge and compass in the obvious way. We now construct a right angled triangle with hypotenuse 4 and adjacent side $1 + \sqrt{5}$ to get an angle of 36°.

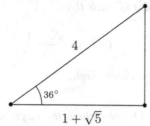

Fig. 6.11   Obtaining an angle of 36°.

Now that we have this angle constructed, we can construct one of the five triangles that make up a regular pentagon, for they have angles of 72°, 54° and 54°, the angle of 72° being at the centre of the circle circumscribing the regular pentagon.

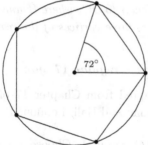

Fig. 6.12   Drawing a regular pentagon.

Problem 6.4. *Show that the following is a method to construct a regular pentagon:* (Solution on page 201.)

Draw a circle, centre O, into which the circle is to be inscribed.
Draw any diameter AOB.
Bisect OB to get C.
Find D on the circle so that OD is perpendicular to AOB.
With centre C and radius CD, draw an arc to intersect AB internally at E.
With centre D and radius DE, draw an arc to intersect the arc of the circle, AD, at F.
Then DF is one side of the regular pentagon required.

Problem 6.5. *Show that the following is a method to construct a regular pentagon:* (Solution on page 202.)

We need to find a point G such that

$$OG = \frac{\sqrt{5} - 1}{4}$$

times the radius of the given circle.

Draw a circle, centre O, into which the circle is to be inscribed.
Let OA be any radius and extend it to B so that $OA = AB$.
Draw BC perpendicular to OB with $BC = OA$.
Note that $OC = \sqrt{5} \times OA$.
Let OC intersect the circle at D.
Find E on the line segment OC such that $CE = OA$.
Bisect OE to find F. Bisect OF to find G.
Draw HGJ perpendicular to OC, where H and J lie on the given circle.
Then H, D and J are three vertices of the regular pentagon.

6.2.4 *Seventeen Sides – regular 17–gon*

This construction is adapted from Chapter 17 of *Galois Theory* by Ian Stewart (2nd Ed., Chapman and Hall, London, 1989).

(1) Draw a circle centre O, radius 1. Choose any point P_0 on the circumference to be the first vertex of the 17–gon.
(2) Draw P_0O and produce it about half as much again.
(3) Draw a line through O at right angles to OP_0 meeting the circle at B.
(4) Find the mid-point J of OB and then the mid-point I of OJ. (The length of OI is $\frac{1}{4}$.)
(5) Bisect $\angle OIP_0$, and then bisect the angle between OI and the bisector of $\angle OIP_0$ to get a line through I meeting OP_0 at E. Then $\angle OIE = \frac{1}{4}\angle OIP_0$.
(6) Draw the line through I a right angles to IE, and bisect the angle between this line and IE to get a line through I meeting P_0O produces at F. Then $\angle EIF = 45°$.
(7) Draw the circle with P_0F as diameter, by first constructing its centre (the mid-point of P_0F — not marked on the diagram). Let K be the point where this circle meets OB.

(8) Draw the circle, centre E, radius EK. This meets OP_0 at N and P_0O produced at M.

(9) Draw lines through N and M perpendicular to OP_0, meeting the original circle at P_3 and P_5.

(10) Draw the circle with centre P_3 and radius P_4P_5 (not shown) to meet the original circle at P_1.

(11) Then $\angle P_0OP_1 = \frac{360}{17}^\circ$. Thus, the 17–gon can be constructed.

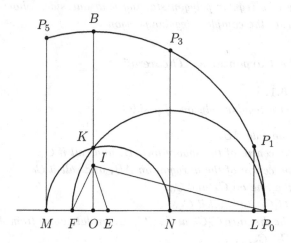

Fig. 6.13 Drawing a regular 17–gon.

What we have here is that the length OL is constructible. (We shall deal with constructions of real numbers later.) It is, in fact, equal to

$$\frac{-1 + \sqrt{17} + \sqrt{34 - 2\sqrt{17}} + \sqrt{68 + 12\sqrt{17} - 16\sqrt{34 + 2\sqrt{17}} - 2\left(1 - \sqrt{17}\right)\sqrt{34 - 2\sqrt{17}}}}{16}$$

See page 173 of Ian Stewart's book, quoted above.

6.2.5 *Other polygons*

The constructions to obtain the other constructible regular polygons are very complicated and beyond the scope of what we do here. However, we will give some "folk theorems" that purport to construct other regular polygons with straight edge and compasses.

WARNING

These results are not to be quoted as being true methods of constructing regular polygons.

First, you should do the following problem:

Problem 6.6. *The centre of a regular polygon is the centre of the circumcircle of that polygon.*

(1) *Show that the centre of a regular polygon is the intersection of the internal bisectors of the interior angles of that polygon.*

(2) *Show how to construct, using straight edge and compasses alone, the centre of a regular polygon starting with one side. Show also how to construct the complete regular polygon.*

6.2.6   *The Carpenter's "Theorem"*

Theorem 6.1.

To construct a regular polygon of n sides:

> *Draw one side AB.*
> *Find the centre of the square on AB, and call it C_4.*
> *Find the centre of the hexagon on AB, and call it C_6.*
> *Draw the line on C_4 and C_6.*
> *Bisect C_4C_6 and call it C_5.*
> *Using the distance C_4C_5 as "unit", extrapolate away from AB on C_4C_6 to get C_7, C_8, ...*
> *The point C_n is the centre of the regular polygon of n sides with side AB.*

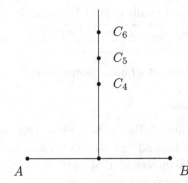

Fig. 6.14   The Carpenter's "Theorem".

This "theorem" occurs, for example, in a book on making furniture. [1]

---

[1] E. Joyce, *The Techniques of Furniture Making*, Batsford, London, 1972

6.2.7 The theory of the Carpenter's "Theorem"

Let C be the mid-point of AB and let $AC = CB = 1$. Then $CC_4 = 1$ and $CC_6 = \sqrt{3}$.

Thus, $C_4C_5 = \dfrac{\sqrt{3}-1}{2}$, so that

$$CC_n = 1 + \frac{(n-4)\left(\sqrt{3}-1\right)}{2} \ .$$

The true centre, C_n^* has the property

$$CC_n^* = \cot\left(\frac{\pi}{n}\right) \ .$$

Thus, the Carpenter's "Theorem" states that

$$\pi = n \tan^{-1}\left(\frac{2}{2+(n-4)\left(\sqrt{3}-1\right)}\right) \ .$$

This is clearly not true, but the question remains as to how good an approximation the right side is to π. We will discuss this later. A graph is shown below to illustrate the way the linear extrapolation used in the Carpenter's "Theorem" (dotted line) approximates the curve on which the centres really lie (solid line).

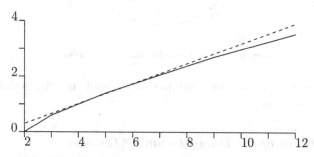

Fig. 6.15   The "theory" of the Carpenter's "Theorem".

6.2.8 The Draughtsman's "Theorem"

Theorem 6.2.

To construct a regular polygon of n sides:

> *Draw the circle.*
> *Draw any diameter AOB.*

Divide AB into n equal portions, and call the points

$$A = P_0, \ P_1, \ P_2, \ \ldots \ P_n = B \ .$$

Find C such that AC = BC = AB.
Draw CP_2 and extend it to meet the circle again at D.
Then AD is a side of the regular polygon.

We give here a diagram to illustrate the construction for a heptagon ($n = 7$).

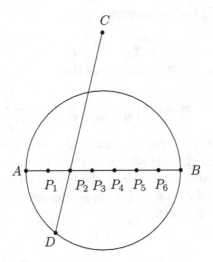

Fig. 6.16   The Draughstsman's "Theorem".

This "theorem" occurs, for example, in a book on the techniques of perspective drawing. [2]

6.2.9   *Theory of the Draughtsman's "Theorem"*

Let O be the origin and $B = (1,0)$. Then $C = (0, \sqrt{3})$ and $P_2 = (4/n - 1, 0)$. For the moment, set $p = 1 - 4/n$. The equation of CP_2 is $y = \sqrt{3}(x - p)/p$, which we solve with $x^2 + y^2 = 1$.
The true position for D^* is given by $(-\cos(2\pi/n), -\sin(2\pi/n))$.
Thus, the Draughtsman's "Theorem" states that

$$\pi = \frac{2}{1 - p} \tan^{-1}\left[\frac{\sqrt{3}}{p}\left(1 - \frac{p^2 + 3}{3 + \sqrt{3 - 2p^2}} \right) \right] \ ,$$

[2]W. Coombes, *Background to Perspective*, A.C. Black, London, 1958

where $p = 1 - 4/n$.

This is clearly not true, but the question remains as to how good an approximation the right side is to π. We will discuss this later.

6.2.10 *The Master Carpenter's "Theorem"*

Theorem 6.3.

The following polygons are easily constructible:

$$\begin{aligned}
\text{Square} \quad & CC_4^* = \cot(\pi/4) = 1; \\
\text{Hexagon} \quad & CC_6^* = \cot(\pi/6) = \sqrt{3}; \\
\text{Octagon} \quad & CC_8^* = \cot(\pi/8) = 1 + \sqrt{2}; \\
\text{Dodecagon} \quad & CC_{12}^* = \cot(\pi/12) = 2 + \sqrt{3}.
\end{aligned}$$

For C_5^*, we use the approximation $CC_5^* = \frac{11}{5} \times AC$ to get

$$1.3764 = \cot(\pi/5) \sim \frac{11}{8} = 1.375.$$

Let CC_7^* be the mid-point of $C_6^* C_8^*$, giving the approximation;

$$2.0765 = \cot(\pi/7) \sim \frac{1 + \sqrt{2} + \sqrt{3}}{2} = 2.0731.$$

Let CC_n^*, ($n = 9, 10, 11$) be even distributed between $C_8^* C_{12}^*$, giving

$$\cot(\pi/n) \sim \frac{4(1 + \sqrt{2}) + (n - 8)(1 - \sqrt{2} + \sqrt{3})}{4}.$$

Written out, we get

$$2.7475 = \cot(\pi/9) \sim \frac{4(1 + \sqrt{2}) + (1 - \sqrt{2} + \sqrt{3})}{4} = 2.7437,$$

$$3.0777 = \cot(\pi/10) \sim \frac{2(1 + \sqrt{2}) + (1 - \sqrt{2} + \sqrt{3})}{2} = 3.0731,$$

$$3.4057 = \cot(\pi/11) \sim \frac{4(1 + \sqrt{2}) + 3(1 - \sqrt{2} + \sqrt{3})}{4} = 3.4026.$$

For C_{13} and beyond, we extrapolate as in the Carpenter's "Theorem", and get better results!

We give a diagram showing this.

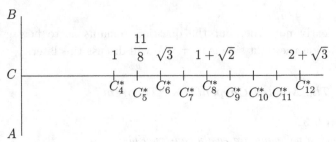

Fig. 6.17   The Master Carpenter's "Theorem".

6.2.11   *Comparison of the three constructions*

The Master Carpenter's "Theorem" is exact in the cases $n = 4$, 6, 8 and 12. The Carpenter's "Theorem" is exact only in the cases $n = 4$ and $n = 6$. The Draughtsman's "Theorem" is never exact!

We shall illustrate the comparison with a short table indicating the accuracy of the constructions.

Method	Number of sides	Approximation to 360°	Percent error	Perimeter error in 48" diameter
Carpenter		361.2	0.57	0.9"
Draughtsman	5	359.2	0.06	−0.1
Master		360.3	0.08	0.1
Carpenter		356.8	0.90	−1.3
Draughtsman	7	360.6	0.17	0.3
Master		360.3	0.08	0.1
Carpenter		350.3	2.70	−4.1
Draughtsman	9	362.5	0.69	1.0
Master		360.5	0.13	0.2
Carpenter		345.0	4.17	−6.3
Draughtsman	11	364.6	1.30	1.9
Master		360.3	0.09	0.1
Carpenter		329.0	8.62	−13.0
Draughtsman	24	365.7	4.20	6.4
Master		355.8	1.16	1.8

6.3 Real Numbers

Here we start with the assumption that we have a line segment of unit length. We ask the question:

> Which line segments can we construct with straight edge and compasses as multiples of the given line segment.

Generally, we shall work with positive real numbers, negative multiples being obtained by reversing the direction.

6.3.1 *Integral Multiples*

Since we can extend a line segment in either direction with straight edge alone, we shall start with the line segment AB (of unit length) and extending in the direction $\overline{AB}$.

$$A \qquad B$$

Fig. 6.18 Integral multiples.

With centre B and radius AB, mark of C on the line on the opposite side of B from A. Then $AC = 2$.

With centre C and the same radius, mark of D on the line on the opposite side of C from B. Then $AD = 3$.

Repeat this process to get any positive integer N.

6.3.2 *Rational Multiples*

At the end of the section entitled Elementary Constructions, we saw how to divide a line segment into m equal parts. Thus, to obtain a rational number p/q, we start with the unit segment, multiply it by p as in the previous subsection, and then divide it into q equal parts.

6.3.3 *Irrational Multiples*

The question of exactly which irrational multiples are possible is beyond the scope of this book. We shall, for the moment, concentrate on numbers of the form $\sqrt{n}$.

The basic result required is Pythagoras's Theorem. We start with a right isosceles triangles.

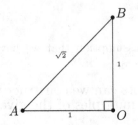

Fig. 6.19   The square root of 2.

Since $OA = OB = 1$, we have $AB^2 = OA^2 + OB^2 = 2$ so that $AB = \sqrt{2}$.
We use this as the basis for proceeding to $\sqrt{3}$, for we now construct a right
triangle with legs 1 and $\sqrt{2}$, to get a hypotenuse of $\sqrt{3}$. And so on.
We illustrate this in the next diagram: we start with the isosceles right
triangle $\triangle OA_1A_2$ which has a right angle at A_1. This gives $OA_2 = \sqrt{2}$.
On OA_2, we construct a right angle at A_2 to give the right triangle
$\triangle OA_2A_3$. This gives $OA_3 = \sqrt{3}$.
And so we proceed, getting $OA_n = \sqrt{n}$, as the "spiral" winds round and
round.

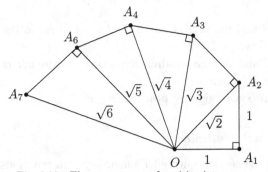

Fig. 6.20   The square root of positive integer n.

Thus, we have shown how to construct $\sqrt{n}$ for any integer n. We can now
find any rational multiple of any square root, and so can find the square
root of any rational number.
We now show how to find the square root of a constructible number p
directly by straight edge and compass construction.
We start with a unit length AB and proceed to find the length that
represents p, calling it BC, where ABC is a line segment. Construct O,
the mid-point of AC, and draw the circle, centred on O, radius OA. Draw
the perpendicular to AC at B, and let it intersect the circle at P and Q.

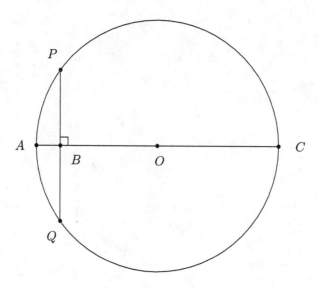

Fig. 6.21 The square root of a constructible number.

Then PB gives $\sqrt{p}$.

We use the intersecting chords theorem to get that

$$AB \cdot BC = PB \cdot BQ \ ,$$

and since $PB = BQ$, we have that $PB = \sqrt{p}$.

Chapter 7

The Broken Chord Theorem

Theorem 7.1. *Let AC and CB be two chords of a circle with $AC < CB$. Let M be the mid-point of that arc AC that passes through C. Let L be the foot of the perpendicular to CB. Then $AC + CL = LB$.*

This theorem is due to Archimedes.

Proof.

We re-write the result as:

Given: a circle with chords AB and BC with $AB < BC$. Let N be the mid-point of AB. The mid-point of the lesser arc AB is M, and L is on BC so that $BC \perp ML$. Extend BC to A' so that $CA' = CA$.

Prove: $AC + CL = LB$.

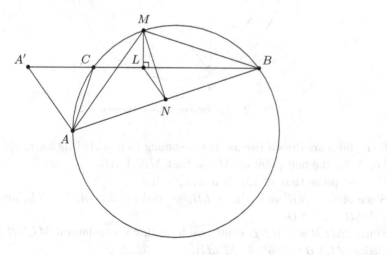

Fig. 7.1   The Broken Chord Theorem – 1.

Let N be the mid-point of AB, so that $MN \perp AB$.

Since $A'C = AC$, we have $\angle A'AC = \angle AA'C = \theta$, say.

Then $\angle ACB = 2\theta$.

Now, $\angle AMB = \angle ACB = 2\theta$ (angles in the same segment).

This means that the circle through A', A and B has centre M.

Now, ML is the perpendicular from the centre, M, to the chord, $A'B$, of this circle.

Hence, L is the mid-point of $A'B$; that is, $A'L = LB$.

Thus, $AC + CL = A'C + CL = A'L = LB$.

As an alternative proof, we have the following:

Given: a circle with chords AB and BC with $AB < BC$. Let N be the mid-point of AB. The mid-point of the lesser arc AB is M, and L is on BC so that $BC \perp ML$. Extend BC to A' so that $BL = LA'$.

Prove: $AC = A'C$.

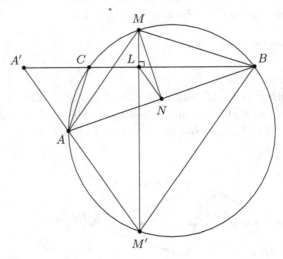

Fig. 7.2   The Broken Chord Theorem – 2.

Some lines are drawn (we are not assuming that $A'AM'$ is a straight line).

Let N be the mid-point of AB, so that $MN \perp AB$.

We first prove that $A'AM'$ is a straight line.

Since $AN = NB$ and $A'L = LB$, we deduce that $AA' \parallel NL$, and that $\angle A'AB = \angle LNB$.

Since $\angle MLB = \angle MNB = 90°$, we have that quadrilateral $MLNB$ cyclic.

Hence, $\angle LNB = 180° - \angle M'MB$.

Since we have quadrilateral $M'AMB$ cyclic, we have $\angle M'MB = \angle M'AB$.

Hence, $\angle A'AB + \angle BAM' = 180°$, so that $A'AM'$ is a straight line.

Since $BL = LA'$ and $M'L \perp BA'$, we now have that $A'M' = M'B$, so that $\triangle A'M'B$ is isosceles.

Thus, $\angle CA'A = \angle CBM'$.

Since we have that quadrilateral $M'ACB$ is cyclic, we deduce that $\angle CBM' = \angle CAA'$.

Thus, $\angle CA'A = \angle CAA'$, and thus, triangle CAA' is isosceles, letting us deduce that $CA = CA'$.

Chapter 8

Cleavers and Splitters

8.1 Cleavers

In N.A. Court's **"College Geometry"**, there is the definition, attributed to D. Avishalom, of a **cleaver**.

Definition: A **cleaver** is a line segment from a mid-point of a side of a triangle that bisects the perimeter of that triangle.

Theorem 8.1. *The cleavers of a triangle are concurrent.*

Lemma 8.1. *Let M be the mid-point of side AC of △ABC. The cleaver, MB', is parallel to the internal angle bisector of ∠ABC.*

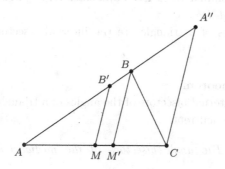

Fig. 8.1   A cleaver.

Proof.
Extend AB to A'' so that $BA'' = BC$. Then $AB' = B'A''$ and $AM = MC$, giving $MB' \parallel CA''$.
As in the "Broken Chord" Theorem, $\angle ABC = 2\angle BA''C$, giving MB' parallel to the angle bisector $M'B$.

119

Suppose that L, M and N are the mid-points of BC, CA and AB, respectively.

Definition: The medial triangle of $\triangle ABC$ is the triangle whose vertices are the mid-points of the sides of $\triangle ABC$.

Lemma 8.2. *The cleaver, MB', bisects $\angle LMN$.*

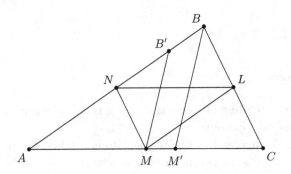

Fig. 8.2 A cleaver as a bisector.

Proof.

Since $\triangle LMN$ is similar to $\triangle ABC$, and since $BM' \parallel MB'$, it follows that MB' bisects $\angle LMN$.

Thus, the cleavers of a triangle are the internal bisectors of the medial triangle $\triangle LMN$.

Proof of the Theorem.

Thus, since the internal bisectors of the angles of a triangle are concurrent, the cleavers are concurrent.

Corollary 8.1. *The angle bisectors of the medial triangle bisect the perimeter.*

Proof.

This is a re-statement of the theorem.

See the diagram in Lemma 8.2.

Definition: The **Spieker Circle** of $\triangle ABC$ is the incircle of the medial triangle. It is named in honour of Theodor Spieker.

Thus, LA', MB' and NC' are cleavers.

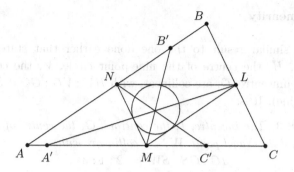

Fig. 8.3 Bisecting the perimeter.

8.2 Splitters

Definition: A **splitter** is a line segment from a vertex to a side of a triangle that bisects the perimeter of that triangle.

Theorem 8.2. *The splitters of a triangle are concurrent.*

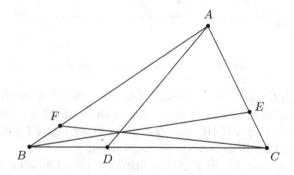

Fig. 8.4 Splitters

If AD is a splitter, then, in the usual notation, $AB = c$ and $BD = s - c$.
Similarly, $AC = b$, $CD = s - b$ and $BD = s - c$.

Thus, $\dfrac{BD}{DC} = \dfrac{s-c}{s-b}$. Similarly, $\dfrac{CF}{FA} = \dfrac{s-a}{s-c}$ and $\dfrac{AE}{EB} = \dfrac{s-b}{s-a}$.

We now use Ceva's Theorem to see that the splitters are concurrent.
The point of concurrency of the splitters is known as the *Nagel Point*.

8.3   Collinearity

There is a similar result to the one done earlier that states that the orthocentre, H, the centre of the nine-point circle, V, the centroid, G, and the circumcentre, O, are collinear, with $HV : VG : GO = 3 : 1 : 2$ (on the Euler Line). It is:

Theorem 8.3. *The incentre, I, the centroid, G, the centre of the Spieker circle, S, and the Nagel point, W, are collinear, with*
$$IG : GS : SW \; = \; 2 : 1 : 3 \; .$$

Proof.
Part (a). We prove that I, G and W are collinear with $IG : GW = 1 : 2$.

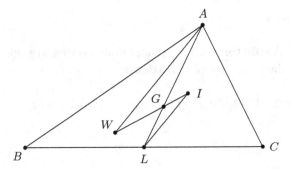

Fig. 8.5   Collinearity of the points – 1.

We start with G and I, and extend IG to W so that $GW = 2IG$.
We recall that $AG = 2GL$. Hence, we have that $\triangle LIG \approx \triangle AWG$. From this, we deduce that $\angle GAW = \angle GLI$, and further that $LI \parallel WA$.
We now show that W is indeed the Nagel point.
Extend AW to meet BC at P. Draw altitude AD and inradius IQ, with D and Q on BC. To show that ALP is a splitter, we show that $AB + BP = s$.
Thus, it is sufficient to show that $BP = s - AB = s - c = z$.
In order to do this, we calculate the length of BD and DP.

Part (a)[i]. The length DP.
Since $AD \parallel IQ$ and $AP \parallel IL$, we have that $\triangle ADP \approx \triangle IQL$. Therefore, $\frac{DP}{QL} = \frac{AD}{IQ}$.
Since $IQ = r$, we have
$$[ABC] \; = \; rs \; = \; \tfrac{1}{2}a(AD) \; .$$

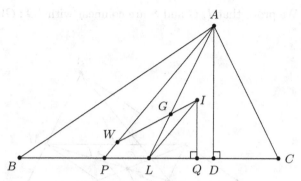

Fig. 8.6   Collinearity of the points – 2.

Combining these results, we obtain that

$$\frac{AD}{IQ} = \frac{AD}{r} = \frac{2s}{a} .$$

Therefore, $DP = QL\frac{2s}{a}$.

Using the Ravi numbers, $x = s - a$, $y = s - b$ and $z = s - c$, we have $BQ = x = s - b$. Hence,

$$QL = BL - BQ = \frac{a}{2} - (s - b) = \frac{b - c}{2} .$$

Thus,

$$DP = \frac{2s}{a}QL = \left(\frac{2s}{a}\right)\left(\frac{b - c}{2}\right) = \frac{s(b - c)}{a} = \frac{(a + b + c)(b - c)}{2a} .$$

Part (a)[ii]. The length BD.

First, we note that $BD = c\cos(B)$. We apply the Cosine Rule to $\triangle ABC$, obtaining

$$b^2 = a^2 + c^2 - 2ac\cos(B) = a^2 + b^2 - 2a\,BD ,$$

so that

$$BD = \frac{a^2 + c^2 - b^2}{2a} .$$

Completion of part (a). Combining these two results, we have

$$BP = BD + DP$$
$$= \frac{a^2 + c^2 - b^2}{2a} + \frac{(a + b + c)(b - c)}{2a}$$
$$= \frac{a^2 + ab - ac}{2a} = \frac{a + b - c}{2} = s - c .$$

Thus, we have that W is indeed the Nagel Point, and that $IG : GW = 1 : 2$.

Part (b). We prove that I, G and S are collinear with $IG : GW = 2 : 1$.

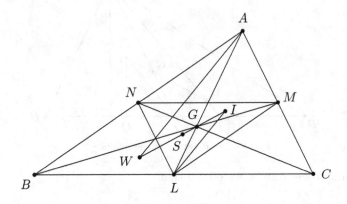

Fig. 8.7   Collinearity of the points – 3.

Consider the medial triangle $\triangle LMN$: S is its incentre, G is its centroid, and I is its Nagel Point (by analogy with $\triangle ABC$).
The proof is complete.

Chapter 9

Remarkable Bisections

In an article [3] in *Mathematics Today* (the journal of the Institute of Mathematics and its Applications), Michael Carding (formerly a high school principal in Cheshire, England) described a visit to India. There, he met a thirteen year old female student who just loved mathematics. She posed him a problem:

In triangle ABC, the perpendiculars from A to the internal bisectors of angles B and C meet those bisectors at X and Y. She claimed that XY is parallel to CB, but had not been able to prove it. She also stated that neither her friends nor her teachers had been able to prove it.

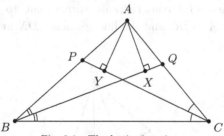

Fig. 9.1 The basic situation.

This was obviously a challenge to Mr. Carding. The article states that Mr. Carding claimed jet-lag and duties to cover his immediate inadequacy, and stated that they have resolved the problem via e-mail. No solution is given in the article.

When first thinking about a problem with bisected angles, it is natural to consider angles! For example, what are the sizes of angles PAY, QAX and XAY? It is also tempting to draw the line segment XY. Also, if $XY \parallel BC$, then we see the Z–theorem coming into play.

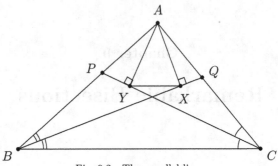

Fig. 9.2   The parallel line.

A little elementary calculation shows us that $\angle PAY = A + \frac{C}{2} - 90°$, that $\angle QAX = A + \frac{B}{2} - 90°$, and that $\angle XAY = \frac{B}{2} + \frac{C}{2}$.

Again, if it were true that $XY \parallel BC$, we would have $\angle AYX = 90° - \frac{C}{2}$ and $\angle AXY = 90° - \frac{B}{2}$. Can we prove these to be true?

However, as in all interesting geometry problems, some construction is necessary. I always told my geometry students (at university particularly) to make as many constructions as seemed to be possibly useful. Later, when a solution has been found, not all of the constructions made will be necessary. Then, a new diagram should be drawn and the old one discarded. In this case, the useful constructions turned out to be drawing the altitude AD from A to BC and the line segments DX and DY.

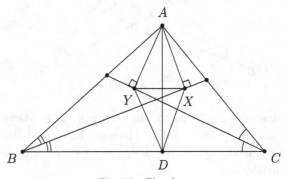

Fig. 9.3   The clue.

From this, we can see that quadrilaterals $AYDC$ and $AXDB$ are cyclic. And this gives lots of equalities amongst angles.

In particular, we have $\angle ADY = \angle ACY = \frac{C}{2} = \angle YCD = \angle YAD$. This tells us that $\triangle AYD$ is isosceles, and thus, $AY = YD$. Similarly, we have $AX = XD$.

This means that quadrilateral $AXDY$ is a very special figure with the property that the diagonals are perpendicular and AD is bisected. Thus, it follows that $XY \parallel BC$, and the theorem is proved. The surprising extra result is that XY bisects the altitude AD. With minor adjustments, this proof is valid for non-acute triangles.

Now, how did I come to think of this proof? To be honest, in preparing a diagram to work with, I resorted to computer algebra. I chose coordinates for A, B and C, worked out the coordinates of the incentre to get the equations of BX and CY. Next, the equations of AX and Y, and then onto the coordinates of X and Y.

I had chosen BC along the x–axis. I was very surprised to discover that the y–coordinates of X and Y were not only equal, but were equal to one half of the y–coordinate of A. This then gave me the clue that I had to prove $AX = XD$ and $AY = YD$. The rest is the proof above.

Now, examining the figure leads to the following result.

Given triangle ABC, let AD be an altitude. Draw the perpendicular bisector of AD and the semicircle ADC. Let Y be the point where these intersect. Then CY is a bisector of $\angle ACD$. This is the internal bisector of $\angle ACB$ if D lies between B and C, and is otherwise the external bisector of $\angle ABC$. We have discovered another way to bisect an angle!

Problem 9.1. *Suppose that L and M are the mid-points of sides BC and AC of triangle ABC, respectively. Suppose that the internal bisector of $\angle ABC$ intersects the line LM at the pont X.*
Prove that CX is perpendicular to BX.

Problem 9.2. *From vertex B of $\triangle ABC$, drop prependiculars to the interior and exterior bisectors of $\angle ABC$, meeting them at P and Q, respectively.*
Prove that PQ bisects both AB and BC.

Problem 9.3. *Suppose that D, E and F are the feet of the altitudes from A, B and C, respectively, in triangle ABC. Prove that H, the othrocentre of triangle ABC, is the incentre of triangle DEF.*

Problem 9.4. *Let H be the orthocentre of* △*ABC. Prove that this diagram is legitimate.* (Solution on page 128.)

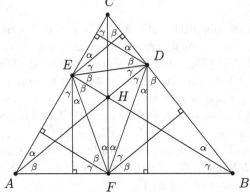

I should like to thank Dr. Chris Fisher of the University of Regina for the following observation.

The 4th edition of *Exercices de Géométrie* written by a Belgian monk, Father Gabriel-Marie (his name was never used on his publications, only his initials), is a 1200-page book, a collection of all geometry theorems known up to 1900, with references to where the results first appeared.

In paragraph 761 on pages 327-328, he has the following result:

The projections from the vertex of a triangle onto the four bisectors of the two other angles lie on a line.

The theorem is attributed to "A. Lascases, de Lorient." (translated as "from Lorient, France".) It appeared in Nouvelles Annales mathématiques de Terquem et Gérono, 1859, p. 171, no. 477. (A later footnote says that Lascases was a student of Gérono.)

Call the first vertex *A* and the other vertices *B* and *C*. The proof by F. G.-M. uses "my student's theorem" that the line in question joins the mid-points of *AB* and *AC*. This claim is easy to see, since the two bisectors of angle *B* form the sides of a rectangle whose other two sides are the perpendiculars dropped from *A*. The side *AB* of triangle *ABC* is then one diagonal of that rectangle, while the other diagonal is the line that joins the feet of the four perpendiculars. The Theorem of Lascases, as well as the above theorem, follow quickly using these rectangles.

Reference

[3] Culture Shock for Mathematics and Science, *Mathematics Today*, Vol. 42, No. 4, 2006, pp. 129–131.

Chapter 10

Miscellaneous Problems

This chapter contains a selection of problems on the material previously covered. The reader must discover which results are required for solutions. Remember that Geometry is an **experimental subject**.

We shall use the notation $[ABC]$ to mean the area of $\triangle ABC$. Similar notation applies to polygons.

Problem 10.1. *In* $\triangle ABC$, *we have* $\angle BAC = 90°$. *Let* D *lie on* BC *so that* $AD \perp BC$. *Suppose that* $AD = 1$.
Show that $BC = AB \times AC$. (See Fig. 10.1.) (Solution on page 201.)

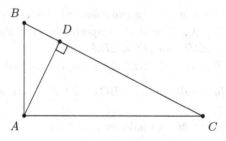

Fig. 10.1

Problem 10.2. *In* $\triangle ABC$, *suppose that* D *is an interior point of the line segment* BC, *that* E *is an interior point of the line segment* CA, *and that* F *is an interior point of the line segment* AB *such that* AD, BE *and* CF *are concurrent.*
Suppose that $\angle BAD = A_1$, $\angle DAC = A_2$, $\angle CBE = B_1$, $\angle EBA = B_2$, $\angle ACF = C_1$ *and* $\angle FCB = C_2$. (See Fig. 10.2.)

Prove that

$$\sin(A_1)\sin(B_1)\sin(C_1) \;=\; \sin(A_2)\sin(B_2)\sin(C_2) \quad.$$

(Solution on page 204.)

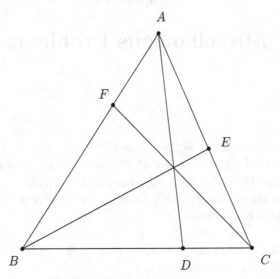

Fig. 10.2   For problem 10.2

Problem 10.3. *Given a convex quadrilateral $ABCD$, extend AB, BC, CD and DA to W, X, Y and Z, respectively such that $AW = 2AB$, $BX = 2BC$, $CY = 2CD$ and $DZ = 2DA$.*
Show that $[WXYZ] = 5 \times [ABCD]$. (Solution on page 206.)

Problem 10.4. *In parallelogram $ABCD$, let E be the mid-point of side AB.*
Prove that DE trisects AC. (Solution on page 206.)

Problem 10.5. *In triangle ABC, let E and D lie on BC such that $CE = ED = DB$. Let F be the mid-point of AC and G be the mid-point of AB. Suppose that EG and FD meet at H.*
Find the value of $\frac{EH}{HG}$. (Solution on page 207.)

Problem 10.6. *On triangle ABC, construct similar isosceles triangles ABC' and CAB' **outwards** on bases AB and CA, respectively, and a further similar isosceles triangle BCA' **inwards** on base BC.*
Prove that $AB'A'C'$ is a parallelogram. (Solution on page 207.)

Problem 10.7. *The sides AB, BC, CD and DA of quadrilateral ABCD are divided by the points E, F, G and H so that*

$$\frac{AE}{EB} = \frac{CF}{FB} = \frac{CG}{GD} = \frac{HA}{DH} .$$

Show that EFGH is a parallelogram. (Solution on page 208.)

Problem 10.8. *In triangle ABC, points D and E lie on sides BC and CA, respectively, so that $\frac{BD}{DC} = 3$ and $\frac{CE}{EA} = \frac{2}{3}$. Suppose that AD and BE meet at P.*
Find the value of $\frac{BP}{PE}$.
Find the value of $\frac{AP}{PD}$. (Solution on page 209.)

Problem 10.9. *In triangle ABC, points E and F lie on sides CA and AB, respectively, so that $\frac{AE}{EC} = 4$ and $\frac{AF}{FB} = 1$. Suppose that D lies on BC and AD meets EF at G. Suppose that $\frac{AG}{GD} = \frac{3}{2}$.*
Find the value of $\frac{BD}{DC}$. (Solution on page 209.)

Problem 10.10. *On the sides of an arbitrary parallelogram ABCD, squares are constructed lying exterior to the parallelogram. Denote the centres of these squares by P_1, P_2, P_3 and P_4.*
Prove that quadrilateral $P_1P_2P_3P_4$ is a square. (Solution on page 210.)

Problem 10.11. *On the sides of an arbitrary convex quadrilateral ABCD, equilateral triangles ABM_1, BCM_2, CDM_3 and DAM_4 are constructed so that ABM_1 and CDM_3 are exterior to the quadrilateral and BCM_2 and DAM_4 are drawn "inwards" to the quadrilateral.*
Show that the quadrilateral $M_1M_2M_3M_4$ is a parallelogram. (Solution on page 211.)

Problem 10.12. *On the sides of an arbitrary convex quadrilateral ABCD, squares are constructed all lying external to the quadrilateral, with centres P_1, P_2, P_3 and P_4.*
Show that $P_1P_3 = P_2P_4$ and that P_1P_3 is perpendicular to P_2P_4. (Solution on page 212.)

Problem 10.13. *Let ABC be an acute angled triangle. Construct squares externally on the sides, and extend the altitudes to meet the opposite sides of the squares, giving rectangles with areas as indicated on the diagram:* (See Fig. 10.3) (Solution on page 213.)
Show that, for i = 1, 2, 3, $A_i = B_i$.

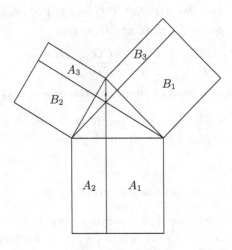

Fig. 10.3

Problem 10.14. *Let A be a given point and let P_i, $i = 1, 2, \ldots, n$, $n \geq 7$, be the vertices of a regular n–gon.*
Let the points Q_i be given by
$$\overrightarrow{AQ_i} \;=\; \overrightarrow{AP_i} \;+\; \overrightarrow{P_{i+1}P_{i+2}} \;, \qquad i = 1, 2, \ldots, n$$
($P_{n+i} = P_i$, $i = 1, 2$).
Prove that the Q_i are the vertices of a regular n–gon. (Solution on page 214.)

Problem 10.15. *Given a circle with centre O and a chord AB, let C be any other point on AB. Join OC and extend it (in the direction $O \to C$) to meet the circle at D.*
Calculate the radius of the circle in terms of AC, BC and CD. (Solution on page 214.)

Problem 10.16. *Given a circle with centre O and two fixed points on the circumference, A and B such that AB is not a diameter.*
Suppose that PQ is any diameter.
Suppose that AP and BQ intersect at X.
Find the locus of X as P moves once around the circle. (Solution on page 215.)

Problem 10.17. *Given a circle and two fixed points on the circumference, A and B, let CD be a chord of fixed length and orientation. Suppose that AC and BD intersect at P.*
Find the locus of P as chord CD moves once around the circle. (Solution on page 216.)

Problem 10.18. *Suppose that $A_0A_1A_2A_3$ is a convex cyclic quadrilateral, and let $A_4 = A_0$.*
Suppose that B_k, $k = 0$, 1, 2 and 3, are points such that

> **arc** A_kB_k = **arc** B_kA_{k+1} $\qquad$ $k = 0, 1, 2, 3$.

Prove that $B_0B_2 \perp B_1B_3$. (Solution on page 218.)

Problem 10.19. *Suppose that $A_0A_1A_2A_3$ is a convex cyclic quadrilateral, and let $A_4 = A_0$.*
Suppose that the diagonals A_0A_2 and A_1A_3 meet at X.
The feet of the perpendicular to A_kA_{k+1} is denoted by B_k, for $k = 0$, 1, 2 and 3. Let $B_4 = B_0$.
Prove that $|B_0B_1| + |B_2B_3| = |B_1B_2| + |B_3B_4|$. (Solution on page 219.)

Problem 10.20. *A chord ST of constant length slides around a semi-circle with diameter AB.*
Let M be the mid-point of ST and P be the foot of the perpendicular from S to AB.
Prove that $\angle SPM$ has the same value, independent of the position of the chord ST. (Solution on page 220.)

Problem 10.21. *Let Γ be a circle and P any fixed point.*
Consider the collection of all lines on P that intersect Γ. Suppose that a typical such line meets the circle at A and B.
Find the locus of the mid-point of AB. (Solution on page 221.)

Problem 10.22. *Suppose that all the angles of $\triangle ABC$ are acute.*
Suppose that AD is an altitude.
Determine the points P and Q thus:

> BP *is parallel to* AC *and* AP *is perpendicular to* AC;
> CQ *is parallel to* AB *and* AQ *is perpendicular to* AB.

Show that AD bisects $\angle PDQ$. (Solution on page 223.)

Problem 10.23. *Find the condition for the straight line $ax + by + c = 0$ to be tangent to the circle, centred at $(0,0)$ and of radius r.* (Solution on page 223.)

Problem 10.24. *In Fig. 10.4, we have two fixed circles, centred on O. A fixed point P lies on the smaller circle. A variable point A also lies on the smaller circle. The chord BC of the larger circle passes through P and is perpendicular to AP.* (Solution on page 224.)

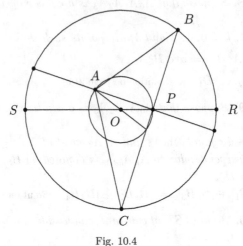

Fig. 10.4

a. Find the set of values of the expression $BC^2 + CA^2 + AB^2$.

b. Find the locus of the mid-point of AB (as A varies around the smaller circle).

Problem 10.25. (See Fig. 10.5.) (Solution on page 226.) *Given triangle ABC and points N_1, N_2 on AB, L_1, L_2 on BC, and M_1, M_2 on CA, respectively, such that*
$$AN_2 = N_2N_1 = N_1B = \frac{AB}{3} \ ,$$

$$BL_2 = L_2L_1 = L_1C = \frac{BC}{3} \ ,$$

$$CM_2 = M_2M_1 = M_1A = \frac{CA}{3} \ .$$
The points A_1, A_2, B_1, B_2, C_1 and C_2 are defined by
$$A_1 = M_1N_1 \cap M_2N_2 \quad A_2 = L_1N_1 \cap L_2M_2$$

$$B_1 = L_1N_1 \cap L_2N_2 \quad B_2 = L_1M_1 \cap N_2M_2$$

$$C_1 = M_1L_1 \cap M_2L_2 \quad C_2 = M_1N_1 \cap L_2N_2 \ .$$

(1) Prove that $L_1M_1 \parallel L_2N_2$.

(2) Prove that $A_1B_1 = A_2B_2$.

(3) Calculate the value of $\frac{[A_1B_1C_1]}{[ABC]}$.

(4) Calculate the value of $\frac{[A_1C_2B_1A_2C_1B_2]}{[ABC]}$.

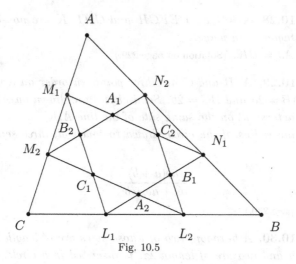

Fig. 10.5

Problem 10.26. *Let AB be a chord of a circle centre O. Let OP be the radius of the circle that passes through the mid-point of the chord AB. Let Q be any point on the circle, on the same side of AB as O.*
Prove that PQ bisects the angle $\angle AQB$. (Solution on page 227.)

Problem 10.27. (Solution on page 227.) *Given triangle ABC, E any point on AC and F any point on AB. Suppose that BE and CF intersect at D.* (See Fig. 10.6.)

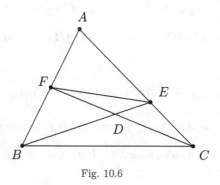

Fig. 10.6

Prove that

$$\frac{[ABC]}{[DBC]} = \frac{[AFE]}{[DFE]}.$$

Problem 10.28. *ABCDE, DEFGH and GHIJK are non-overlapping regular pentagons in a plane.*
Prove that $AJ = BK$. (Solution on page 229.)

Problem 10.29. *A, B and C are three points in order on a straight line such that $AB = 2a$ and $BC = 2b$. Semi-circles are drawn on AB, BC and CA as diameters, all on the same side of the line ABC.*
Show that the radius of the circle drawn to touch all three semi-circles is given by

$$\frac{ab(a+b)}{a^2 + ab + b^2} \ .$$

(Solution on page 230.)

Problem 10.30. *A hexagon, two of whose sides are of length $2x$, two are of length $2y$ and two are of length $2z$, is inscribed in a circle. Prove that the radius ρ of this circle is given by the equation*

$$\rho^3 - (x^2 + y^2 + z^2)\rho = 2xyz \ .$$

(Solution on page 231.)

Problem 10.31. (Solution on page 232.) *Let O be the centre of $\triangle ABC$. Points A_B, A_C, B_C, B_A, C_B and C_A all lie on the circumcircle and satisfy;*

$$\angle ABA_B = \angle A_BBO \ , \qquad\qquad \angle OBC_B = \angle C_BBC \ ,$$

$$\angle BCB_C = \angle B_CCO \ , \qquad\qquad \angle OCA_C = \angle A_CCA \ ,$$

$$\angle CAC_A = \angle C_AAO \ , \qquad\qquad \angle OAB_A = \angle B_AAB \ .$$

Prove that

$$\text{Arc } A_CB_C + \text{Arc } B_AC_A + \text{Arc } C_BA_B$$

$$= \text{Arc } A_BA_C + \text{Arc } B_CB_A + \text{Arc } C_AC_B \ .$$

Problem 10.32. (Solution on page 233.) *If a, b, c are the sides of a triangle with area Δ, show that*

$$4\sqrt{3}\Delta \ \leq \ a^2 + b^2 + c^2 \ .$$

Problem 10.33. *Suppose that A_i, $i = 1, 2, 3, 4$ are the interior angles of a convex quadrilateral (so that $A_1 + A_2 + A_3 + A_4 = 2\pi$ and $0 < A_i < \pi$). Show that $\prod_{i=1}^{4} \sin\left(\frac{A_i}{2}\right) \leq \frac{1}{4}$.* (Solution on page 234.)

Problem 10.34. *Suppose that A_i, $i = 1, 2, 3, 4$ are the interior angles of a convex quadrilateral (so that $A_1 + A_2 + A_3 + A_4 = 2\pi$ and $0 < A_i < \pi$). Show that $\prod_{i=1}^{4} \cos\left(\frac{A_i}{2}\right) \leq \frac{1}{4}$.* (Solution on page 234.)

Problem 10.35. *Suppose that A_i, $i = 1, 2, 3$ are the interior angles of an acute angled triangle. Show that $\prod_{i=1}^{3} \tan(A_i) \geq 3\sqrt{3}$.* (Solution on page 235.)

Problem 10.36. *Suppose that $\angle A + \angle B + \angle C = 180°$. Prove that*

$$\begin{vmatrix} 1 & 1 & 1 \\ \cos(A) & \cos(B) & \cos(C) \\ \tan\left(\frac{A}{2}\right) & \tan\left(\frac{B}{2}\right) & \tan\left(\frac{C}{2}\right) \end{vmatrix} = 0 \ .$$

(Solution on page 235.)

Problem 10.37. *Suppose that $\angle A + \angle B + \angle C = 180°$. Prove that*

$$\frac{1}{\sin(A/2)} + \frac{1}{\sin(B/2)} + \frac{1}{\sin(C/2)} \geq 6 \ .$$

(Solution on page 237.)

Problem 10.38. *Suppose that R and r represent the circumradius and inradius of $\triangle ABC$. Prove that*

$$\frac{2R}{r} = \frac{\sin(A) + \sin(B) + \sin(C)}{\sin(A)\sin(B)\sin(C)} \geq 4 \ .$$

Prove also that

$$\frac{\sin(A) + \sin(B) + \sin(C)}{\sin(A)\sin(B)\sin(C)} = 4$$

if and only if $A = B = C = 60°$. (Solution on page 238.)

Problem 10.39. *Consider a trapezoid $ABCD$ with $AB \parallel CD$. Let P be the point of intersection of the diagonals AC and BD.* (See Fig. 10.7.) (Solution on page 238.)
Choose any point M on AB and produce MP to meet DC at N.
Let AN and MD meet at Q. Let BN and MC meet at R.
Prove that P, Q and R lie on a line.

Problem 10.40. (Solution on page 239.) *In an acute angled triangle ABC with orthocentre H and circumcentre O, show that*

$$\angle HAO = |\angle ABC - \angle ACB| \ .$$

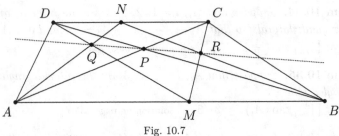

Fig. 10.7

Problem 10.41. *Let AD, BE and CF be the internal bisectors of the angles of ABC, with D, E and F on BC, CA and AB, respectively.*
If quadrilateral BCEF is cyclic, show that ABC is isosceles. (Solution on page 240.)

Problem 10.42. *Triangle ABC has circumcentre O. The mid-points of AB, BC and CA are F, D and E, respectively.*
Produce DO, EO and FO to meet EF, FD and DE at D', E' and F', respectively.
Show that O is the incentre of D'E'F'. (Solution on page 240.)

Problem 10.43. *In triangle ABC, let D be the point on BC such that AB + BD = AC + CD. Similarly, let E and F be on AC and AB, respectively, such that BA + AE = BC + CE and CA + AF = CB + BF. Show that AD, BE and CF are concurrent.* (Solution on page 241.)

Problem 10.44. *Hexagon ABCDEF is such that a circle can be inscribed to touch all six sides.*
Show that AB + CD + EF = BC + DE + FA. (Solution on page 242.)

Problem 10.45. *In triangle ABC, let D, E and F be the mid-points of BC, CA and AB, respectively. Let D', E' and F' be the reflections of D, E and F in the angle bisectors of ∠CAB, ∠ABC and ∠BCA, respectively. Show that AD', BE' and CF' are concurrent.* (Solution on page 242.)

Problem 10.46. *In triangle ABC, the points D, E and F lie on BC, CA and AB, respectively, and are such that BC = 3BD, CA = 3CE and AB = 3AF.*
Show that triangles ABC and DEF have the same centroid. (Solution on page 243.)

Problem 10.47. *Isosceles triangles BCX, CAY and ABZ are constructed externally to the sides of triangle ABC.*
Show that triangles ABC and XYZ have the same centroid. (Solution on page 244.)

Problem 10.48. *For triangle ABC, denote the lengths for the altitudes from A, B and C by h_a, h_b and h_c, respectively.*
Extend the altitudes at A, B and C through the opposite sides to A', B' and C', respectively, where $AA' = \frac{k}{h_a}$, $BB' = \frac{k}{h_b}$ and $CC' = \frac{k}{h_c}$, k being a positive constant.
Prove that triangles ABC and $A'B'C'$ have the same centroid. (Solution on page 244.)

Problem 10.49. *In $\triangle ABC$, let O be the circumcentre. Draw perpendiculars from O to each side: $OD \perp BC$, $OE \perp CA$ and $OF \perp AB$.*
Extend each to D', E' and F', respectively, so that $OD = DD'$, $OE = EE'$ and $OF = FF'$.
Prove that $\triangle D'E'F'$ is congruent to $\triangle ABD$. (Solution on page 245.)

Problem 10.50. *Let C be a fixed circle and T a tangent line to C.*
Let C_0 be a circle, externally tangent to C and with T as a tangent line.
Define a sequence of circles: $\{C_n\}_{n=1}^{\infty}$ by:

a. C_n is externally tangent to C and to C_{n-1}; and
b. T is a tangent line to C_n

Find the locus of the centre of C_{1999} as the radius of C_0 is allowed to vary in size. (Solution on page 246.)

Problem 10.51. (Solution on page 247.) *By using an appropriate placement of a triangle in the Cartesian coordinate system, prove the following:*

(a) the medians of a triangle are concurrent.
(b) the altitudes of a triangle are concurrent.

Problem 10.52. (Solution on page 248.) *Prove that the altitudes of a triangle are concurrent:*

a. using Ceva's theorem;
b. using only properties of triangles and circles, but NOT Ceva's theorem or Menelaus' theorem.

Problem 10.53. *Suppose that Γ is a fixed circle and AB a fixed chord of that circle.*
Let C be any other point on Γ. Let G be the centroid of $\triangle ABC$.
Find the locus of G as C varies around the circle Γ. (Solution on page 249.)

Problem 10.54. *Suppose that Γ is a fixed circle and AB a fixed chord of that circle.*
Let C be any other point on Γ. Let H be the orthocentre of $\triangle ABC$.
Find the locus of H as C varies around the circle Γ. (Solution on page 249.)

Problem 10.55. (Solution on page 250.) *Given $\triangle ABC$ with orthocentre H,*
let Γ be the nine-point circle and Λ the circumcircle.
(a) Show that the radius of Λ is twice the radius of Γ.
Let P be any point on Λ. Construct the **line segment** *PH and let it intersect Γ at Q.*
(b) Prove that $HQ = QP$.

Problem 10.56. *In $\triangle ABC$, the points C_A, C_1 and C_B lie on the line segment AB such that $AC_A = C_A C_1 = C_1 C_B = C_B B$. Similarly, the points B_A, B_1 and B_C lie on the line segment AC such that $AB_A = B_A B_1 = B_1 B_C = B_C C$, and the points A_B, A_1 and A_C lie on the line segment BC such that $BA_B = A_B A_1 = A_1 A_C = A_C C$.*
Show that $\triangle ABC$, $\triangle A_B B_C C_A$ and $\triangle A_C C_B B_A$ all have the same centroid.
(Solution on page 251.)

Problem 10.57. *Triangle ABC has sides $AB = 39cm$, $BC = 42cm$ and $CA = 45cm$. Points E and F lie on the line segments AC and AB, respectively, so that*

$$AE = \frac{AC}{3} \quad and \quad BF = \frac{BA}{3} ,$$

respectively. The line EF meets the line BC at D.
Calculate the length of the line segment BD. (Solution on page 251.)

Problem 10.58. *Triangle ABC is inscribed in a circle Γ. The internal bisectors of the angles $\angle CAB$, $\angle ABC$ and $\angle BCA$ meet Γ again at U, V and W, respectively.*
Perpendiculars for U, V and W meet the sides BC, CA and AB, respectively, at L, M and N, respectively.
Prove that AL, BM and CN are concurrent. (Solution on page 252.)

Problem 10.59. *Triangle XYZ has sides $XY = 24$, $YZ = 16$ and $ZX = 20$. Points E and F lie on the line segments XZ and XY, respectively, so that*

$$XE = \frac{XZ}{4} \quad and \quad YF = \frac{YX}{4},$$

respectively. The line EF meets the line YZ at D.
Calculate the length of the line segment DY. (Solution on page 253.)

Problem 10.60. *Triangle PQR is inscribed in a circle Λ. U, V and W are the mid-points of QR, RP and PQ, respectively.*
Draw perpendiculars to the sides of the triangle at U, V and W outwards away from the interior of the triangle, to meet the circle Λ at the points L, M and N, respectively.
Prove that PL, QM and RN are concurrent. (Solution on page 254.)

Problem 10.61. *Given quadrilateral $ABCD$ and a transverse line $PQRS$, where P lies on AB, Q lies on BD, R lies on AC and S lies on DC (none being coincident with any vertex of the quadrilateral).* (See Fig. 10.8.)
(Solution on page 254.)

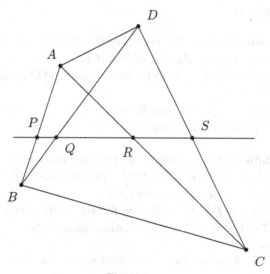

Fig. 10.8

Show that

$$\frac{AP}{PB} \cdot \frac{DS}{SC} = \frac{AR}{RC} \cdot \frac{DQ}{QB} \ .$$

Problem 10.62. (Solution on page 255.)

   a. *Given triangle $\triangle ABC$, let D, E and F be the mid-points of BC, CA*
      and AB, respectively.
      The circle, centre E and radius EC, intersects the circle, centre D and
      radius DC, at C and P.
      Show that P lies on AB.
   b. *Similarly, find Q on BC and R on CA.*
      Show that AQ, BR and CP are concurrent.
   c. *Determine what this point of concurrency is.*

Problem 10.63. *Suppose that I_A, I_B and I_C are the centres of the escribed circles of the triangle ABC, opposite to A, B and C, respectively. Show that the centre of the inscribed circle of the triangle ABC is the orthocentre of the triangle $I_A I_B I_C$.* (Solution on page 256.)

Problem 10.64. *The inscribed circle of triangle ABC touches the sides BC, CA and AB in X, Y and Z, respectively. I is the centre of the circle. If XI intersects ZY in L, prove that AL is a median of triangle ABC.* (Solution on page 256.)

Problem 10.65. *Let O be the centre of the circumcircle Γ of a regular convex polygon $A_1 A_2 A_3 \ldots A_{25} A_{26}$ of 26 sides.*
Construct the points O_1 and O_2 to be the reflections of O in the lines $A_{25} A_1$ and $A_2 A_6$, respectively.
Prove that $O_1 O_2$ is equal in length to a side of an equilateral triangle inscribed in Γ. (Solution on page 258.)

Problem 10.66. *Let us consider any **non-equilateral** triangle with circumcentre O, centroid G, in-centre I and orthocentre H.*
Let V be the centre of the nine point circle.
Let P be the foot of the perpendicular from I to the Euler line $OGVH$.
Prove that P lies between G and H. (Solution on page 259.)

Problem 10.67. *Suppose that Γ is a fixed circle and AB a fixed chord of that circle.*
Let C be any other point on Γ. Let J be the centre of the nine point circle of $\triangle ABC$.
Find the locus of J as C varies around the circle Γ. (Solution on page 260.)

Problem 10.68. *In* $\triangle ABC$, *let* D, E *and* F *be points in the interiors of the line segments* BC, CA *and* AB, *respectively. Suppose that* AD, BE *and* CF *are concurrent at* P. *Let* X, Y *and* Z *be the mid-points of* PA, PB *and* PC, *respectively.*
Show that the circumradius of $\triangle XYZ$ *is half the circumradius of* $\triangle ABC$.
(Solution on page 260.)

Problem 10.69. *Suppose that the sides of right* $\triangle ABC$ *are all integers. Prove that the inradius is an integer.* (Solution on page 261.)

Problem 10.70. *Suppose that* Γ *is a fixed circle and* AB *a fixed chord of that circle.*
Let C *be any other point on* Γ. *Let* I *be the incentre of* $\triangle ABC$.
Find the locus of I *as* C *varies around the circle* Γ. (Solution on page 261.)

Problem 10.71. *Suppose that* Γ *is a fixed circle and* AB *a fixed chord of that circle.*
Let C *be any other point on* Γ. *Let* I_A, I_B *and* I_C *be the centres of the escribed circles of* $\triangle ABC$.
Find the locus of I_A, I_B *and* I_C *as* C *varies around the circle* Γ. (Solution on page 262.)

Problem 10.72. *Obtain the following for* **cyclic quadrilaterals** *with sides*
$AB = a$, $BC = b$, $CD = c$, $DA = d$ *and diagonals* $AC = e$, $BD = f$, *and where* $s = \frac{a+b+c+d}{2}$ *and* R *is the radius of the circle:* (Solution on page 263.)

(1) $[ABCD] = \sqrt{(s-a)(s-b)(s-c)(s-d)}$.
(2) $R = \frac{1}{4}\sqrt{\frac{(ab+cd)(ac+bd)(bc+ad)}{(s-a)(s-b)(s-c)(s-d)}}$.
(3) $e = \sqrt{\frac{(ac+bd)(bc+ad)}{ab+cd}}$.

Problem 10.73. *Suppose that quadrilateral* $ABCD$ *has an incircle of radius* r.
If $AB = a$, $BC = b$, $CD = c$, $DA = d$, *and* $s = \frac{a+b+c+d}{2}$, *prove that*
(Solution on page 265.)

(1) $a + c = b + d$;
(2) $[ABCD] = rs$.

Problem 10.74. *Show that*
$$s\tan\left(\frac{A}{2}\right) + 2R\cos(A) = 2R + r \ .$$

Problem 10.75. *Show that, if the sides of a triangle are in arithmetic progression, then*

$$2s^2 = 9Rr - 18r^2 \ .$$

(Solution on page 267.)

Problem 10.76. *Show that, if the sides of a triangle are in geometric progression, then*

$$\left(s^2 + 4Rr + r^2\right)^3 = 32Rrs^4 \ .$$

(Solution on page 268.)

Problem 10.77. *Prove the following equalities about the angles of* $\triangle ABC$:
(Solution on page 268.)

(1) $\sin(A) + \sin(B) + \sin(C) = \frac{s}{R}$,
(2) $\cos(A) + \cos(B) + \cos(C) = \frac{R+r}{R}$,
(3) $\cos(A)\cos(B) + \cos(B)\cos(C) + \cos(C)\cos(A) = \frac{r^2 + S^2 - 4r^2}{4R^2}$,
(4) $\cos(A)\cos(B)\cos(C) = \frac{s^2 - (2R+r)^2}{4R^2}$,
(5) $\tan(A) + \tan(B) + \tan(C) = \frac{2rs}{s^2 - 4R^2 - 4Rr - r^2} = \frac{2rs}{s^2 - (2R+r)^2}$,
(6) $\tan(A/2) + \tan(B/2) + \tan(C/2) = \frac{4R+r}{s}$.

Problem 10.78. *H is the orthocentre of ABC. Given the points A, B and H, find a straight edge and compass construction to find the point C.*
(Solution on page 269.)

Problem 10.79. *Show how to construct triangle with sides equal to and parallel to the medians of a given triangle.* (Solution on page 270.)

Problem 10.80. *Let V be the vertex of a parabola, let C be an arbitrary point on the parabola, and let M be the mid-point of VC. Prove that the locus of M is a parabola as C varies on the original parabola. Find the focus, directrix and vertex of this new parabola.*

Problem 10.81. *Let V be the vertex of a parabola with focus F, and let m be the line through V, parallel to the directrix. Let M and N be two distinct points (different from V) on the parabola such that M, F and N are collinear. Let P and Q be on m such that* $MP \perp m$ *and* $NQ \perp m$. *Prove that* $MP \cdot NQ$ *is constant.*

Problem 10.82. *Find the eccentricity, foci and directrices of each of the following conic sections:*

(a) $\frac{x^2}{25} + \frac{y^2}{9} = 1$; (b) $\frac{x^2}{4} + \frac{y^2}{9} = 1$;

(c) $\frac{x^2}{25} - \frac{y^2}{9} = 1$; (d) $\frac{y^2}{25} - \frac{x^2}{9} = 1$; (e) $xy = 9$.

Problem 10.83. *Let A and B be the end-points of the major axis of an ellipse. Let P be any point on the ellipse, distinct from A and B. Show that the product of the slopes of the lines PA and PB is constant.*

Problem 10.84. *Let Γ be a circle and AB a fixed diameter of Γ. Let CD be any chord of the circle perpendicular to AB. Let P and Q be the points on the line segment CD such that $CP = PQ = QD$.*
Prove that the loci of P and Q is an ellipse (as CD varies), and determine the foci of the ellipse (in terms of A and B).

Problem 10.85. *A straight line intersects a hyperbola at the points K and L, and the hyperbola's asymptotes at R and T. Prove that the mid-point of KL is also the mid-point of RT.*

Problem 10.86. *Consider the hyperbolas with equations $\frac{x^2}{a^2} - \frac{y^2}{b^2} = 1$ and $\frac{x^2}{b^2} - \frac{y^2}{a^2} = 1$. (Such hyperbolas are called conjugate.)*
If e_1 and e_2 are the eccentricities of these hyperbolas, show that $e_1^2 + e_2^2 = e_1^2 e_2^2$.

Problem 10.87. *Let AB be a fixed line segment, let C be a point not on AB, and let m be the length of the line segment joining C to the mid-point of AB.*
Show that the locus of all points C satisfying the equation $m^2 = AC \cdot BC$ is a hyperbola.

Problem 10.88. (Solution on page 271.) *Give a purely geometric proof that*

$$\tan^{-1}\frac{1}{2} + \tan^{-1}\frac{1}{3} = \frac{\pi}{4} \ .$$

Problem 10.89. (Solution on page 272.) *On the planet, Nobis, bisection of line segments by straight edge and compasses is forbidden.*
You are commanded by King Youclid to start with line segments of lengths 16 and 12, and, with straight edge and compasses, construct a line segment of length 7.
Can you obey the king's command?

Problem 10.90. (Solution on page 272.) *Rectangle $ABCD$ has $AB = \frac{BC}{2}$.*
Outside the triangle, draw $\triangle DCF$, where $\angle DFC = 30°$ and ADF is a
straight line segment. Let E be the mid-point of AD.
Determine the measure of $\angle EBF$.

Problem 10.91. (Solution on page 273.) *A square of side length s is inscribed*
symmetrically inside a sector of a circle with radius of length r and central
angle of $60°$, such that two vertices lie on the straight sides of the sector
and two vertices lie on the circular arc of the sector.
Determine the exact value of $\frac{s}{r}$.

Problem 10.92. (Solution on page 274.) *The points A_0, A_1, ..., A_n lie on a*
line, ordered from left to right. The circles Γ_k on $A_{k-1}A_k$ $(k = 1, ..., n)$
as diameters all have the same radius r, and have centres O_k, respectively.
A tangent line is drawn from A_0 to Γ_n, intersecting the circle Γ_k at the
points B_k and C_k. Here, $B_1 = A_0$, $B_n = C_n = P$, the point of tangency
on Γ_n, and the point B_k is to the left of the point C_k.
Determine the length of the line segment B_kC_k, where this tangent
intersects the circle Γ_k.

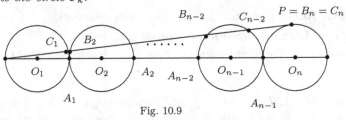

Fig. 10.9

Problem 10.93. (Solution on page 275.) *The line with slope $\lambda > 0$ acts like*
a mirror to a ray of light coming along a line parallel to the x–axis from
$+\infty$. Determine the slope of the reflected ray.

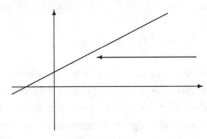

Fig. 10.10

Problem 10.94. (Solution on page 276.) *Given triangle ABC and DE ∥ BC, with D ∈ AB and E ∈ AC. Drop perpendiculars from D and E to BC, meeting BC at F and K, respectively.*
If $\frac{[ABC]}{[DEKF]} = \frac{32}{7}$, *determine the ratio* $\frac{|AD|}{|DB|}$.

Problem 10.95. (Solution on page 277.) *Determine the maximum area of an isosceles triangle inscribed in a circle of radius r.*

Problem 10.96. (Solution on page 277.) *Square ABCD is inscribed in one eighth of a circle of radius 1, such that there is one vertex on each radius and two vertices on the arc.*

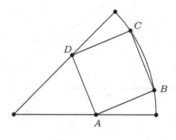

Fig. 10.11

Determine the area of the square. Exact answer required in the form $\frac{a+b\sqrt{c}}{d}$, *where a, b, c and d are integers.*

Problem 10.97. (Solution on page 278.) *A right angled triangle ABC has the following property:*
draw a square externally on each side; the vertices of the squares not on the right triangle are concyclic.
Characterize such triangles.

Problem 10.98. (Solution on page 280.) *In triangle ABC, angle BAC is a right angle and AC > AB. The mid-points of BC, AC, and AB are L, M, and N, respectively.*
The circle, centre L, radius $\frac{AB}{2}$ intersects BL at P and CL at Q.
If PN ∥ QM, determine the size of angle ACB.

Problem 10.99. (Solution on page 281.) *In square ABCD is the mid-point of AC, F is the point on B such that DF ⊥ EB, and G is the point on EF such that AG ⊥ EB.*
Show that CF = DG.

Problem 10.100. (Solution on page 282.)
$\triangle ABC$ has a right angle at B, and I is its incentre. Quadrilateral $ADEB$ is a rectangle with $I \in DE$. Quadrilateral $CGFB$ is a rectangle with $I \in FG$. Let $H = DE \cap AC$, and $K = FG \cap AC$.
Prove that

$$\triangle HIK = \triangle ADH + \triangle CGK .$$

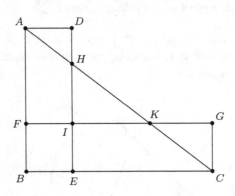

Fig. 10.12

Problem 10.101. *Triangle ABC with $a = BC \geq CA = b$ has a right angle at C. Squares $ABDE$, $BCFG$ and $CAHI$ are drawn externally to triangle ABC as shown in Fig. 10.13.* (Solution on page 282.)
Let $FI \cap EH = P$, $IF \cap DG = Q$ and $GD \cap HE = R$.

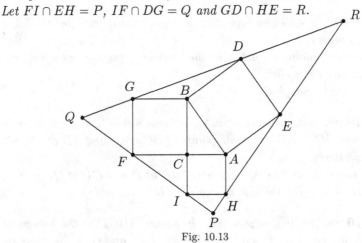

Fig. 10.13

Determine the ratio $\frac{b}{a}$ such that triangle PQR is a right triangle.

Problem 10.102. (Solution on page 283.) *On a certain day, the moon is seen with the shadow passing through diametrically opposite points. A photograph is shown.*

Fig. 10.14

Calculate the exact proportion of the moon that is bright as seen on the photograph Fig. 10.14.

HINT: Remember that the shadow of a sphere appears as a circle.

Problem 10.103. *Given a circle of radius 6 and circles of radii 4, 4 and 2, each of which intersect the circle of radius 6, but not each other, creating seven regions with areas A, B, C, D, E, F, and G as shown in Fig. 10.15.*

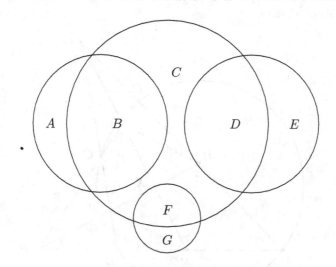

Fig. 10.15

Determine the exact value of $A + E + G - C$.

Problem 10.104. *Two squares $XZBA$ and $YZCD$ are drawn outside $\triangle XYZ$.*

Prove that the point of intersection of XD and YA lies on the altitude of $\triangle XYZ$ which passes through the vertex Z. (Solution on page 285.)

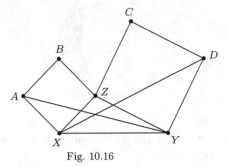

Fig. 10.16

Problem 10.105. *Given cyclic quadrilateral $ABCD$, suppose that the lines AB and CD meet at P, and that the lines AD and BC meet at Q. The internal bisector of $\angle AQB$ meets DC and AB at G and E, respectively. The internal bisector of $\angle APD$ meets BC and AD at F and H, respectively.* (Solution on page 286.)

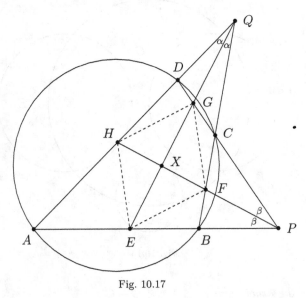

Fig. 10.17

Prove that $EFGH$ is a rhombus.

Problem 10.106. *The sides AB, BC, DC and AD of quadrilateral ABCD are extended to E, F, G and H, respectively, and have the property that AE, BF, CG and DH are tangent to a circle at E, F, G and H, respectively.* (Solution on page 288.)

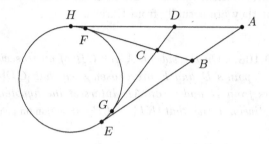

Prove that $AB + BC = AD + DC$.

Problem 10.107. *The internal bisectors of angles $\angle ACB$ and $\angle CAB$ meet the opposite sides AB and CB at E and D, respectively.*
Let F be an point on the line segment ED. Drop perpendiculars FG, FH and FN to sides AB, BC and CA, respectively. (Solution on page 288.)

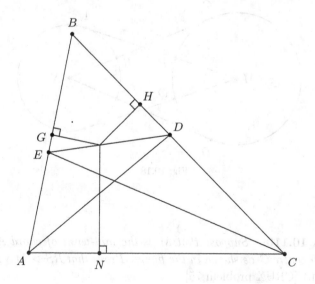

Prove that $FG + FH = FN$.

10.1   Problems from Crux Mathematicorum with Mathematical Mayhem

These problems are reproduced here with the kind permission of the Canadian Mathematical Society. Those from Crux are from volumes 1 and 2; so that they are actually from Eureka.

Problem 10.108. *On the sides CA and CB of an isosceles right-angled triangle ABC, points D and E are chosen such that $|CD| = |CE|$. The perpendiculars from D and C on AE intersect the hypotenuse AB in K and L, respectively. Prove that $|KL| = |LB|$.* (Solution on page 290.) [CRUX problem 33]

Problem 10.109. *From the centres of each of two non-intersecting circles tangents are drawn to the other circle, as shown in the diagram below. Prove that the chords PQ and RS are equal in length. (It is reputed that this problem originated with Newton, but I have not been able to find an exact reference.)* (Solution on page 291.) [CRUX problem 63]

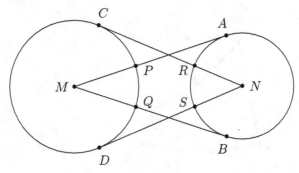

Fig. 10.18

Problem 10.110. *Suppose that M is the mid-point of chord AB of the circle with centre C as shown in the figure. Prove that $RS > MN$.* (Solution on page 291.) [CRUX problem 75]

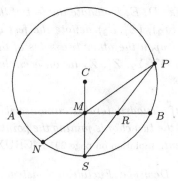

Fig. 10.19

Problem 10.111. *Show that, for any triangle ABC,*

$$|OA|^2 \sin A + |OB|^2 \sin B + |OC|^2 \sin C = 2K \ ,$$

where O is the centre of the inscribed circle and K is the area of $\triangle ABC$.
(Solution on page 292.) [CRUX problem 126]

Problem 10.112. *In square ABCD,* $\overline{AC}$ *and* $\overline{BD}$ *meet at E. Point F is in* $\overline{CD}$ *and* $\angle CAF = \angle FAD$. *If* $\overline{AF}$ *meets* $\overline{ED}$ *at G and if EG = 24, find CF.* (Solution on page 293.) [CRUX problem 147]

Problem 10.113. *Consider the isosceles triangle ABC in the figure, which has a vertical angle of* $20°$. (Solution on page 294.) [CRUX problem 175]

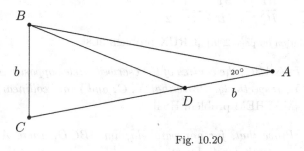

Fig. 10.20

On AC, one of the equal sides, a point D is marked off so that $|AD| = |BC| = b$. *Find the measure of* $\angle ABD$.

Problem 10.114. *Let D, E, F denote the feet of the altitudes of $\triangle ABC$, and let (X_1, X_2), (Y_1, Y_2), (Z_1, Z_2) denote the feet of perpendiculars from D, E, F, respectively, upon the other two sides of the triangle. Prove that the six points X_1, X_2, Y_1, Y_2, Z_1, Z_2 lie on a circle.* (Solution on page 295.) [CRUX problem 192]

Problem 10.115. *If a quadrilateral is circumscribed about a circle, prove that its diagonals and the two chords joining the points of contact of opposite sides are all concurrent.* (Solution on page 297.) [CRUX problem 199]

Problem 10.116. *Devise a Euclidean construction to divide a given line segment into two parts such that the sum of the squares on the whole segment and on one of its parts is equal to twice the square on the other part.* (Solution on page 298.) [CRUX problem 158]

Problem 10.117. *If a, b, c are the sides of a triangle ABC, t_a, t_b, t_c are the angle bisectors, and T_a, T_b, T_c are the angle bisectors extended until they are chords of the circle circumscribing the triangle ABC, prove that* (Solution on page 298.) [CRUX problem 168]

$$abc = \sqrt{T_a T_b T_c t_a t_b t_c} \ .$$

Problem 10.118. *A square $PQRS$ is inscribed in a semicircle (O) with PQ falling along diameter AB (see figure). A right triangle ABC, equivalent to the square, is inscribed in the same semicircle with C lying on the arc RB. Show that the incentre I of triangle ABC lies at the intersection of SB and RQ, and that*

$$\frac{RI}{IQ} = \frac{SI}{IB} = \frac{1 + \sqrt{5}}{2} \ ,$$

the golden ratio. (Solution on page 299.) [CRUX problem 368]

Problem 10.119. *In $\triangle ABC$, the centres of the escribed circles opposite A and B are at X and Y, respectively. Prove that X, C, and Y are collinear.* (Solution on page 300.) [MAYHEM problem HS4]

Problem 10.120. *Prove that, for any parallelogram $ABCD$, where A, B, C, and D are consecutively labelled vertices,*

$$AB^2 + BC^2 + CD^2 + DA^2 = AC^2 = BD^2.$$

(Solution on page 301.) [MAYHEM problem HS6]

Problem 10.121.   *In $\triangle ABC$, the bisector of $\angle A$ intersects BC at D. Prove that*

$$AD^2 \; = \; AB \cdot AC - BD \cdot CD.$$

(Solution on page 302.) [MAYHEM problem HS17]

Problem 10.122.   *In $\triangle ABC$, $AB = AC$. If D is on BC extended such that $BC = CD$ and E is on AB extended, so that $AB = BE$, show that $AD = CE$.* (Solution on page 303.) [MAYHEM problem HS20]

Problem 10.123.   *Let D and E be points on sides AC and AB respectively of $\triangle ABC$. If $AD = AE$ and $BD = CE$, prove that $AB = AC$.* (Solution on page 303.) [MAYHEM problem HS33]

Problem 10.124.   *Two circles C_1 and C_2, with centres O_1 and O_2, are externally tangent. PQ is a common tangent to the circles, with P on C_1 and Q on C_2, and M is the mid-point of PQ. Show that $\angle O_1 M O_2 = 90°$.* (Solution on page 304.) [MAYHEM problem HS64]

Chapter 11

Problems with Solutions

Most of these problems have appeared before in this book. But, for clarity, we repeat the statement before giving the solution.

Problem 11.1. (This is Problem 1.2 on page 14.) *In an equilateral triangle* $\triangle ABC$, *of side* 2, M *and* N *are the mid-points of* AB *and* AC, *respectively. The triangle is inscribed in a circle. The line segment* MN *is extended to meet the circle at* P. *Find the length of the line segment* NP.

Solution. Extend the chord PNM to meet the circle again at Q.

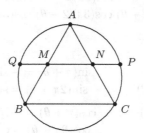

Fig. 11.1   For problem 11.1.

Note that $PN = QM$ (by symmetry). Denote their common length by λ. We also have $AN = NC = MN = 1$. By the Intersecting Chords Theorem, we have $QN.NP = AN.NC$, so that $(1 + \lambda)\lambda = 1$.
The solutions to this are $\lambda = \frac{-1 \pm \sqrt{5}}{2}$. But $\lambda > 0$. Hence, $\lambda = \frac{\sqrt{5}-1}{2}$.

Problem 11.2. (This is Problem 2.1 on page 22.) *Prove the following equalities:*

(1) $\sin^2(\theta) + \cos^2(\theta) = 1$;
(2) $\tan^2(\theta) - \sec^2(\theta) = -1$;
(3) $\cot^2(\theta) - \csc^2(\theta) = -1$.

157

Solution. We shall use Pythagoras' Theorem in right triangle ABC with angle θ.

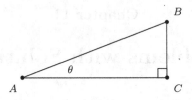

Fig. 11.2   For Problem 11.2.

(1) $\sin^2(\theta) + \cos^2(\theta) = \dfrac{BC^2}{AB^2} + \dfrac{AC^2}{AB^2} = \dfrac{BC^2 + AC^2}{AB^2} = 1$;

(2) $\tan^2(\theta) - \sec^2(\theta) = \dfrac{\sin^2(\theta)}{\cos^2(\theta)} - \dfrac{1}{\cos^2(\theta)} = \dfrac{\sin^2(\theta) - 1}{\cos^2(\theta)} = -1$;

(3) $\cot^2(\theta) - \csc^2(\theta) = \dfrac{\cos^2(\theta)}{\sin^2(\theta)} - \dfrac{1}{\sin^2(\theta)} = \dfrac{\cos^2(\theta) - 1}{\sin^2(\theta)} = -1$.

Problem 11.3. (This is Problem 2.2 on page 25.) *Give the following in the form:* $\pm \sin(\theta)$ *or* $\pm \cos(\theta)$:

$$\sin(\pi - \theta) \quad \sin(\pi + \theta) \quad \sin(2\pi - \theta) \quad \sin(2\pi + \theta)$$
$$\cos(\pi - \theta) \quad \cos(\pi + \theta) \quad \cos(2\pi - \theta) \quad \cos(2\pi + \theta)$$
$$\sin(\pi/2 - \theta) \; \sin(\pi/2 + \theta) \; \sin(3\pi/2 - \theta) \; \sin(3\pi/2 + \theta)$$
$$\cos(\pi/2 - \theta) \; \cos(\pi/2 + \theta) \; \cos(3\pi/2 - \theta) \; \cos(3\pi/2 + \theta) \quad .$$

Solution.

$$\sin(\pi - \theta) = \sin(\theta) \qquad\qquad \sin(\pi + \theta) = -\sin(\theta)$$
$$\sin(2\pi - \theta) = -\sin(\theta) \qquad\qquad \sin(2\pi + \theta) = \sin(\theta)$$

$$\cos(\pi - \theta) = -\cos(\theta) \qquad\qquad \cos(\pi + \theta) = -\cos(\theta)$$
$$\cos(2\pi - \theta) = \cos(\theta) \qquad\qquad \cos(2\pi + \theta) = \cos(\theta)$$

$$\sin(\pi/2 - \theta) = \cos(\theta) \qquad\qquad \sin(\pi/2 + \theta) = \cos(\theta)$$
$$\sin(3\pi/2 - \theta) = -\cos(\theta) \; \sin(3\pi/2 + \theta) = -\cos(\theta)$$

$$\cos(\pi/2 - \theta) = \sin(\theta) \qquad \cos(\pi/2 + \theta) = -\sin(\theta)$$
$$\cos(3\pi/2 - \theta) = -\sin(\theta) \quad \cos(3\pi/2 + \theta) = \sin(\theta) \quad .$$

Problem 11.4. (This is Problem 2.3 on page 25.) *Give the following in the form:* $\pm \tan(\theta)$ *or* $\pm \cot(\theta)$:

$$\tan(\pi - \theta) \quad \tan(\pi + \theta) \quad \tan(2\pi - \theta) \quad \tan(2\pi + \theta)$$
$$\cot(\pi - \theta) \quad \cot(\pi + \theta) \quad \cot(2\pi - \theta) \quad \cot(2\pi + \theta)$$
$$\tan(\pi/2 - \theta) \; \tan(\pi/2 + \theta) \; \tan(3\pi/2 - \theta) \; \tan(3\pi/2 + \theta)$$
$$\cot(\pi/2 - \theta) \; \cot(\pi/2 + \theta) \; \cot(3\pi/2 - \theta) \; \cot(3\pi/2 + \theta) \quad .$$

Solution.

$$\begin{aligned}
\tan(\pi - \theta) &= -\tan(\theta) & \tan(\pi + \theta) &= \tan(\theta) \\
\tan(2\pi - \theta) &= -\tan(\theta) & \tan(2\pi + \theta) &= \tan(\theta)
\end{aligned}$$

$$\begin{aligned}
\cot(\pi - \theta) &= -\cot(\theta) & \cot(\pi + \theta) &= \cot(\theta) \\
\cot(2\pi - \theta) &= -\cot(\theta) & \cot(2\pi + \theta) &= \cot(\theta)
\end{aligned}$$

$$\begin{aligned}
\tan(\pi/2 - \theta) &= \cot(\theta) & \tan(\pi/2 + \theta) &= -\cot(\theta) \\
\tan(3\pi/2 - \theta) &= \cot(\theta) & \tan(3\pi/2 + \theta) &= -\cot(\theta)
\end{aligned}$$

$$\begin{aligned}
\cot(\pi/2 - \theta) &= \tan(\theta) & \cot(\pi/2 + \theta) &= -\tan(\theta) \\
\cot(3\pi/2 - \theta) &= \tan(\theta) & \cot(3\pi/2 + \theta) &= -\tan(\theta) \quad .
\end{aligned}$$

Problem 11.5. (This is Problem 2.4 on page 25.) *Give the following in terms of* $\sin(\theta)$, $\cos(\theta)$ *or* $\tan(\theta)$:

$$\begin{aligned}
&\sin(-\theta) & &\cos(-\theta) & &\tan(-\theta) & &\cot(-\theta) \\
&\sin(2\pi + \theta) & &\cos(2\pi + \theta) & &\tan(2\pi + \theta) & &\cot(2\pi + \theta) \\
&\sin(3\pi + \theta) & &\cos(4\pi + \theta) & &\tan(5\pi + \theta) & &\cot(6\pi + \theta) \\
&\sin(3\pi - \theta) & &\cos(3\pi/2 - \theta) & &\tan(7\pi - \theta) & &\cot(9\pi/2 - \theta) \quad .
\end{aligned}$$

Solution.

$$\begin{aligned}
\sin(-\theta) &= -\sin(\theta) & \cos(-\theta) &= \cos(\theta) \\
\tan(-\theta) &= -\tan(t) & \cot(-\theta) &= \tfrac{-1}{\tan(\theta)}
\end{aligned}$$

$$\begin{aligned}
\sin(2\pi + \theta) &= \sin(\theta) & \cos(2\pi + \theta) &= \cos(\theta) \\
\tan(2\pi + \theta) &= \tan(t) & \cot(2\pi + \theta) &= \tfrac{1}{\tan(\theta)}
\end{aligned}$$

$$\begin{aligned}
\sin(3\pi + \theta) &= -\sin(\theta) & \cos(4\pi + \theta) &= \cos(\theta) \\
\tan(5\pi + \theta) &= \tan(\theta) & \cot(6\pi + \theta) &= \tfrac{1}{\tan(\theta)}
\end{aligned}$$

$$\begin{aligned}
\sin(3\pi - \theta) &= \sin(\theta) & \cos(3\pi/2 - \theta) &= -\sin(\theta) \\
\tan(7\pi - \theta) &= \tfrac{1}{\tan(t)} & \cot(9\pi/2 - \theta) &= \tan(\theta) \quad .
\end{aligned}$$

Problem 11.6. (This is Problem 2.5 on page 27.) *Prove the following equalities:*

(1) $\sin(\theta - \phi) = \sin(\theta)\cos(\phi) - \cos(\theta)\sin(\phi)$;

(2) $\cos(\theta + \phi) = \cos(\theta)\cos(\phi) - \sin(\theta)\sin(\phi)$;

(3) $\cos(\theta - \phi) = \cos(\theta)\cos(\phi) + \sin(\theta)\sin(\phi)$;

(4) $\tan(\theta + \phi) = \frac{\tan(\theta)+\tan(\phi)}{1-\tan(\theta)\tan(\phi)}$;

(5) $\tan(\theta - \phi) = \frac{\tan(\theta)-\tan(\phi)}{1+\tan(\theta)\tan(\phi)}$.

Solution. (1) $\sin(\theta - \phi) = \sin(\theta)\cos(\phi) - \cos(\theta)\sin(\phi)$;

This is obtained by replacing ϕ by $-\phi$ in $\sin(\theta + \phi) = \sin(\theta)\cos(\phi) + \cos(\theta)\sin(\phi)$.

(2) $\cos(\theta + \phi) = \cos(\theta)\cos(\phi) - \sin(\theta)\sin(\phi)$;

We use the figure as in the text:

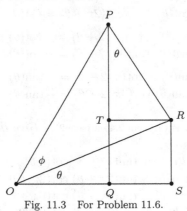

Fig. 11.3   For Problem 11.6.

We are given $\angle SOR = \theta$ and $\angle ROP = \phi$.

Let $OP = r$, $RT = SQ = p$, $OR = w$, $TP = s$, $OQ = q$ and $PR = v$. Note that $\angle OQP = \angle ORP = \pi/2$, so that quadrilateral $OQRP$ is cyclic. Thus, $\angle QRP = \angle ROS = \theta$.

We have

$$\sin(\theta) = \frac{TP}{RP} = \frac{s}{v} \qquad \cos(\theta) = \frac{OS}{OR} = \frac{q+p}{w}$$

$$\sin(\phi) = \frac{PR}{PO} = \frac{v}{r} \qquad \cos(\phi) = \frac{OR}{OP} = \frac{w}{r}$$

so that

$$\cos(\theta)\cos(\phi) - \sin(\theta)\sin(\phi) = \frac{q+p}{w}\cdot\frac{w}{r} - \frac{s}{v}\cdot\frac{v}{r} = \frac{q}{r}$$

$$= \frac{OQ}{PO} = \cos(\theta + \phi) \ .$$

(3) $\cos(\theta - \phi) = \cos(\theta)\cos(\phi) + \sin(\theta)\sin(\phi)$;

This is obtained from the previous result by replacing ϕ by $-\phi$.

(4) $\tan(\theta + \phi) = \frac{\tan(\theta)+\tan(\phi)}{1-\tan(\theta)\tan(\phi)}$.

We note that

$$\tan(\theta + \phi) = \frac{\sin(\theta + \phi)}{\cos(\theta + \phi)} = \frac{\sin(\theta)\cos(\phi) + \cos(\theta)\sin(\phi)}{\cos(\theta)\cos(\phi) - \sin(\theta)\sin(\phi)} \ .$$

Dividing top and bottom by $\cos(\theta)\cos(\phi)$ leads to the result.

(5) $\tan(\theta - \phi) = \frac{\tan(\theta) - \tan(\phi)}{1 + \tan(\theta)\tan(\phi)}$.

This is obtained from the previous result by replacing ϕ by $-\phi$.

Problem 11.7. (This is Problem 2.6 on page 27) *Find other proofs (of the double angle formulae) by searching in books.*

Solution. Did you search the literature?

Problem 11.8. (This is Problem 2.7 on page 27) *Find formulae for*
$$\cos(A) + \cos(B), \quad \cos(A) - \cos(B) .$$
What can you do for $\tan(A) + \tan(B)$?

Solution.
$$\cos(A) + \cos(B) = 2\cos\left(\frac{A+B}{2}\right)\cos\left(\frac{A-B}{2}\right)$$

$$\cos(A) - \cos(B) = 2\sin\left(\frac{A+B}{2}\right)\sin\left(\frac{B-A}{2}\right)$$

For $\tan(A) + \tan(B)$, not much can be done meaningfully!

Problem 11.9. (This is Problem 2.8 on page 29.) *Solve the following trigonometric equations:*

(1) $(a^2 - b^2)\sin(\theta) + 2ab\cos(\theta) = (a^2 + b^2)$.
(2) $\sin(\theta) + \cos(\theta) = 3$.
(3) $\sin(\theta) - \cos(\theta) = \sqrt{2}$.

Solution. We use the half-angle t substitutions:

(1)
$$\begin{aligned}
0 &= -(a^2 - b^2)\sin(\theta) - 2ab\cos(\theta) + (a^2 + b^2) \\
&= -(a^2 - b^2)\frac{2t}{1+t^2} - 2ab\frac{1-t^2}{1+t^2} + (a^2 + b^2) \\
&= \frac{-2t(a^2 - b^2) - 2ab(1 - t^2) + (a^2 + b^2)(1 + t^2)}{1 + t^2} \\
&= \frac{t^2(a^2 + 2ab + b^2) - 2t(a^2 - b^2) + (a^2 - 2ab + b^2)}{1 + t^2} \\
&= \frac{t^2(a + b)^2 - 2t(a - b)(a + b) + (a - b)^2}{1 + t^2} \\
&= \frac{\left((a + b)t + (a - b)\right)^2}{1 + t^2}
\end{aligned}$$

so that $t = \tan\left(\frac{\theta}{2}\right) = \frac{a-b}{a+b}$.

We need specific values to solve this!

(2) $\sin(\theta) + \cos(\theta) = 3$.

There is no solution since $\sin(\theta) + \cos(\theta)$ cannot exceed 2.

(3) $\sin(\theta) - \cos(\theta) = \sqrt{2}$.

We show two solutions, one involving half angle formulae, the other direct.

(a)

$$\sin(\theta) - \cos(\theta) = \sqrt{2} \longleftrightarrow \frac{2t}{1+t^2} - \frac{1-t^2}{1+t^2} = \sqrt{2}$$

$$\longleftrightarrow \frac{2t - 1 + t^2}{1+t^2} = \sqrt{2}$$

$$\longleftrightarrow t^2 + 2t - 1 = \sqrt{2}(1+t^2)$$

$$\longleftrightarrow t^2(\sqrt{2} - 1) - 2t + (\sqrt{2} + 1) = 0$$

$$\longleftrightarrow t^2 - 2(\sqrt{2} + 1)t + (\sqrt{2} + 1)^2 = 0 \ ,$$

so that $t = \tan\left(\frac{\theta}{2}\right) = \sqrt{2} + 1$.

To solve this, we find $\sin(\theta) = \frac{2t}{1+t^2} = \frac{2(\sqrt{2}+1)}{1+(3+2\sqrt{2})} = \frac{2(\sqrt{2}+1)}{2\sqrt{2}(\sqrt{2}+1)} = \frac{1}{\sqrt{2}}$.

Also, $\cos(\theta) = \frac{1-t^2}{1+t^2} = \frac{1-(3+2\sqrt{2})}{1+(3+2\sqrt{2})} = \frac{-1}{\sqrt{2}}$.

(We now know that θ is in the second quadrant.)

Thus, $\sin(2\theta) = 2\sin(\theta)\cos(\theta) = -1$, from which we see that $\theta = \frac{3\pi}{4}$, etc.

(b) We re-write $\sin(\theta) - \cos(\theta) = \sqrt{2}$ as

$$\sin(\theta)\frac{1}{\sqrt{2}} + \cos(\theta)\frac{-1}{\sqrt{2}} = 1 \ .$$

Now, we note that $\frac{1}{\sqrt{2}} = \cos(7\pi/4)$ and that $\frac{-1}{\sqrt{2}} = \sin(7\pi/4)$. Thus we have

$$\sin(\theta)\cos(7\pi/4) + \cos(\theta)\sin(7\pi/4) = \sin(\theta + 7\pi/4) = 1 \ .$$

Thus, $\theta + 7\pi/4 = \pi/2$, etc.

Problem 11.10. (This is Problem 2.9 on page 29.) *Prove the following, where A, B, C are the angles of a triangle.*

(1) $\sin(A) + \sin(B) + \sin(C) = 4\cos\left(\frac{A}{2}\right)\cos\left(\frac{B}{2}\right)\cos\left(\frac{C}{2}\right)$;

(2) $\cos(A) + \cos(B) + \cos(C) = 1 + 4\sin\left(\frac{A}{2}\right)\sin\left(\frac{B}{2}\right)\sin\left(\frac{C}{2}\right)$;

(3) $\sin(2A) + \sin(2B) + \sin(2C) = 4\sin(A)\sin(B)\sin(C)$;

(4) $\cos(2A) + \cos(2B) + \cos(2C) = -\big(1 + 4\cos(A)\cos(B)\cos(C)\big)$;

(5) $\tan(A) + \tan(B) + \tan(C) = \tan(A)\tan(B)\tan(C)$;

(6) $\sin^2(A) + \sin^2(B) + \sin^2(C) = 2\big(1 + \cos(A)\cos(B)\cos(C)\big)$;

(7) $\cos^2(A) + \cos^2(B) + \cos^2(C) = 1 - 2\cos(A)\cos(B)\cos(C)$;

(8) $\cot(A)\cot(B) + \cot(B)\cot(C) + \cot(C)\cot(A) = 1$;

(9) $\cot(\frac{A}{2}) + \cot(\frac{B}{2}) + \cot(\frac{C}{2}) = \cot(\frac{A}{2})\cot(\frac{B}{2})\cot(\frac{C}{2})$;

(10) $\big(\sin(A) + \sin(B) + \sin(C)\big) \times \big(-\sin(A) + \sin(B) + \sin(C)\big)$
$\times \big(\sin(A) - \sin(B) + \sin(C)\big) \times \big(\sin(A) + \sin(B) - \sin(C)\big)$
$= 4\sin^2(A)\sin^2(B)\sin^2(C)$.

Solution.

(1)

$$\sin(A) + \sin(B) + \sin(C)$$
$$= \sin(A) + \sin(B) + \sin(A+B)$$
$$= 2\sin\left(\frac{A+B}{2}\right)\cos\left(\frac{A-B}{2}\right) + 2\sin\left(\frac{A+B}{2}\right)\cos\left(\frac{A+B}{2}\right)$$
$$= 2\sin\left(\frac{A+B}{2}\right)\left(\cos\left(\frac{A+B}{2}\right) + \cos\left(\frac{A-B}{2}\right)\right)$$
$$= 2\cos(\tfrac{C}{2})\left(2\cos(\tfrac{A}{2})\cos(\tfrac{B}{2})\right) = 4\cos(\tfrac{A}{2})\cos(\tfrac{B}{2})\cos(\tfrac{C}{2})$$.

(2)

$$\cos(A) + \cos(B) + \cos(C)$$
$$= \cos(A) + \cos(B) + 1 - 2\sin^2(\tfrac{C}{2})$$
$$= 2\cos\left(\frac{A+B}{2}\right)\cos\left(\frac{A-B}{2}\right) + 1 - 2\cos^2\left(\frac{A+B}{2}\right)$$
$$= 1 + 2\cos\left(\frac{A+B}{2}\right)\left(\cos\left(\frac{A-B}{2}\right) - \cos\left(\frac{A+B}{2}\right)\right)$$
$$= 1 + 2\cos\left(\frac{A+B}{2}\right) 2\sin(\tfrac{A}{2})\sin(\tfrac{B}{2}) = 1 + 4\sin(\tfrac{A}{2})\sin(\tfrac{B}{2})\sin(\tfrac{C}{2})$$.

(3)
$$\sin(2A) + \sin(2B) + \sin(2C)$$
$$= 2\sin(A+B)\sin(A-B) + 2\sin(C)\cos(C)$$
$$= 2\sin(C)\cos(A-B) - 2\sin(C)\cos(A+B)$$
$$= 2\sin(C)\big(\cos(A-B) - \cos(A+B)\big)$$
$$= 2\sin(C)\,2\sin(A)\sin(B) \;=\; 4\sin(A)\sin(B)\sin(C) \;\;.$$

(4)
$$\cos(2A) + \cos(2B) + \cos(2C)$$
$$= 2\cos(A+B)\cos(A-B) + 2\cos^2(C) - 1$$
$$= -2\cos(C)\cos(A-B) - 1 - 2\cos(C)\cos(A+B)$$
$$= -1 - 2\cos(C)\big(\cos(A-B) + \cos(A+B)\big)$$
$$= -1 - 2\cos(C)\,2\cos(A)\cos(B) \;=\; -\big(1 + 4\cos(A)\cos(B)\cos(C)\big) \;\;.$$

(5)
$$\tan(A) + \tan(B) + \tan(C)$$
$$= \tan(A) + \tan(B) - \tan(A+B)$$
$$= \tan(A) + \tan(B) - \frac{\tan(A) + \tan(B)}{1 - \tan(A)\tan(B)}$$
$$= \big(\tan(A) + \tan(B)\big)\left(1 - \frac{1}{1 - \tan(A)\tan(B)}\right)$$
$$= \big(\tan(A) + \tan(B)\big)\left(\frac{-\big(\tan(A)\tan(B)\big)}{1 - \tan(A)\tan(B)}\right).$$
$$= -\tan(A)\tan(B)\left(\frac{\tan(A) + \tan(B)}{1 - \tan(A)\tan(B)}\right)$$
$$= -\tan(A)\tan(B)\tan(A+B)$$
$$= \tan(A)\tan(B)\tan(C) \;\;.$$

(6)
$$\sin^2(A) + \sin^2(B) + \sin^2(C)$$
$$= -\frac{1}{2}\big(1 - 2\sin^2(A) + 1 - 2\sin^2(B) + 1 - 2\sin^2(C) - 3\big)$$
$$= \frac{1}{2}\Big(3 - \big(\cos(2A) + \cos(2B) + \cos(2C)\big)\Big)$$
$$= \frac{1}{2}\big(3 + 1 + 4\cos(A)\cos(B)\cos(C)\big)$$
$$= 2\big(1 + \cos(A)\cos(B)\cos(C)\big) \;\;.$$

(7)

$$\cos^2(A) + \cos^2(B) + \cos^2(C)$$
$$= \frac{1}{2} \left(2\cos^2(A) - 1 + 2\cos^2(B) - 1 + 2\cos^2(C) - 1 + 3 \right)$$
$$= \frac{1}{2} \left(\cos(2A) + \cos(2B) + \cos(2C) \right)$$
$$= \left(3 - 1 - 4\cos(A)\cos(B)\cos(C) \right)$$
$$= 1 - 2\cos(A)\cos(B)\cos(C) \ .$$

(8)

$$\cot(A)\cot(B) + \cot(B)\cot(C) + \cot(C)\cot(A)$$
$$= \frac{\cos(A)\cos(B)}{\sin(A)\sin(B)} + \frac{\cos(B)\cos(C)}{\sin(B)\sin(C)} + \frac{\cos(C)\cos(A)}{\sin(C)\sin(A)}$$
$$= \frac{\cos(A)\cos(B)\sin(C)}{\sin(A)\sin(B)\sin(C)} + \frac{\sin(A)\cos(B)\cos(C)}{\sin(A)\sin(B)\sin(C)} + \frac{\cos(A)\sin(B)\cos(C)}{\sin(A)\sin(B)\sin(C)}$$
$$= \frac{\cos(A)\cos(B)\sin(A+B)}{\sin(A)\sin(B)\sin(C)} - \frac{\sin(A)\cos(B)\cos(A+B)}{\sin(A)\sin(B)\sin(C)}$$
$$\qquad - \frac{\cos(A)\sin(B)\cos(A+B)}{\sin(A)\sin(B)\sin(C)}$$
$$= \frac{\cos(A)\cos(B)\sin(A+B) - \cos(A+B)\sin(A+B)}{\sin(A)\sin(B)\sin(C)}$$
$$= \frac{\sin(A+B)\ \sin(A)\sin(B)}{\sin(A)\sin(B)\sin(C)}$$
$$= \frac{\sin(A)\sin(B)\sin(C)}{\sin(A)\sin(B)\sin(C)} = 1 \ .$$

(9) Here we need both:

$$\tan(\theta + \phi) = \frac{\cot(\theta) + \cot(\phi)}{\cot(\theta)\cot(\phi) - 1} \quad \text{and} \quad \cot(\tfrac{\pi}{2} - \theta) = \tan(\theta) \ .$$

Thus:

$$\cot(\tfrac{A}{2}) + \cot(\tfrac{B}{2}) + \cot(\tfrac{C}{2})$$
$$= \cot(\tfrac{A}{2}) + \cot(\tfrac{B}{2}) + \tan\left(\tfrac{A+B}{2}\right)$$
$$= \left(\cot(\tfrac{A}{2}) + \cot(\tfrac{B}{2})\right) + \left(\frac{\cot(\tfrac{A}{2}) + \cot(\tfrac{B}{2})}{\cot(\tfrac{A}{2})\cot(\tfrac{B}{2}) - 1}\right)$$
$$= \left(\cot(\tfrac{A}{2}) + \cot(\tfrac{B}{2})\right)\left(\frac{\cot(\tfrac{A}{2})\cot(\tfrac{B}{2})}{\cot(\tfrac{A}{2})\cot(\tfrac{B}{2}) - 1}\right)$$
$$= \tan\left(\frac{A+B}{2}\right)\cot(\tfrac{A}{2})\cot(\tfrac{B}{2})$$
$$= \cot(\tfrac{A}{2})\cot(\tfrac{B}{2})\cot(\tfrac{C}{2}) \ .$$

(10) We deal with each part separately:
$$\left(\sin(A) + \sin(B) + \sin(C)\right)$$
$$\times \left(-\sin(A) + \sin(B) + \sin(C)\right)$$
$$\times \left(\sin(A) - \sin(B) + \sin(C)\right)$$
$$\times \left(\sin(A) + \sin(B) - \sin(C)\right)$$
From the first part on this problem, we have:

$$\sin(A) + \sin(B) + \sin(C) = 4\cos(\tfrac{A}{2})\cos(\tfrac{B}{2})\cos(\tfrac{C}{2}) \ .$$

Using the same techniques, we have:

$$\sin(A) + \sin(B) - \sin(C)$$
$$= 2\sin\left(\frac{A+B}{2}\right)\left(\cos\left(\frac{A+B}{2}\right) - \cos\left(\frac{A-B}{2}\right)\right)$$
$$= 2\cos(\tfrac{C}{2})\, 2\sin(\tfrac{A}{2})\sin(\tfrac{B}{2}) \ .$$

This gives that the expression is equal to:

$$64\sin^2(\tfrac{A}{2})\cos^2(\tfrac{A}{2})\sin^2(\tfrac{B}{2})\cos^2(\tfrac{B}{2})\sin^2(\tfrac{C}{2})\cos^2(\tfrac{C}{2})$$
$$= 4\sin^2(A)\sin^2(B)\sin^2(C) \ .$$

Problem 11.11. (This is Problem 2.11 on page 32.) *Find different proofs of the Sine Rule.*

Solution. How many did you find?

Problem 11.12. (This is Problem 2.12 on page 32.) *Prove* **Ptolemy's Theorem**: *that is, if ABCD is a cyclic quadrilateral, then we have* $AC \cdot BD = AB \cdot CD + AD \cdot BC$.

Solution. We name some angles in the diagram:

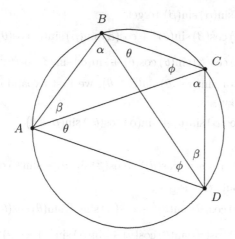

Fig. 11.4   For Problem 11.12.

First, note that $\alpha + \beta + \theta + \phi = 180°$.
Now, rewrite the required result as

$$\frac{AB}{AC} \cdot \frac{CD}{BD} + \frac{AD}{AC} \cdot \frac{BC}{BD} = 1 \ .$$

Using the Sine Rule, this is the same as

$$\frac{\sin(\theta)}{\sin(\phi + \alpha)} \cdot \frac{\sin(\phi)}{\sin(\theta + \alpha)} + \frac{\sin(\alpha)}{\sin(\theta + \beta)} \cdot \frac{\sin(\beta)}{\sin(\theta + \alpha)} = 1 \ .$$

Since $\phi + \alpha = 180° - \theta - \beta$, both denominators are the same.
Also, $\sin(\phi) = \sin(180° - \alpha - \beta - \theta) = \sin(\alpha + \beta + \theta)$.
Thus, what we have to prove is the same as

$$\frac{\sin(\theta)\sin(\alpha + \beta + \theta) + \sin(\alpha)\sin(\beta)}{\sin(\alpha + \theta)\sin(\beta + \theta)} = 1 \ .$$

Note that $\sin(\alpha + \beta + \theta)$ can be expanded to get

$$\cos(\alpha)(\cos(\beta)\sin(\theta) + \sin(\beta)\cos(\theta)) + \sin(\alpha)(\cos(\beta)\cos(\theta) - \sin(\beta)\sin(\theta)) \ .$$

Further expansion gives

$$\cos(\alpha)\cos(\beta)\sin(\theta) + \cos(\alpha)\sin(\beta)\cos(\theta)$$

$$+ \sin(\alpha)\cos(\beta)\cos(\theta) - \sin(\alpha)\sin(\beta)\sin(\theta) \ ,$$

and, when multiplied by $\sin(\theta)$, we get

$$\cos(\alpha)\cos(\beta)\sin(\theta)^2 + \cos(\alpha)\sin(\beta)\sin(\theta)\cos(\theta)$$

$$+ \sin(\alpha)\cos(\beta)\sin(\theta)\cos(\theta) - \sin(\alpha)\sin(\beta)\sin(\theta)^2 \ .$$

We must now add $\sin(\alpha)\sin(\beta)$ to get

$$\cos(\alpha)\cos(\beta)\sin(\theta)^2 + \cos(\alpha)\sin(\beta)\sin(\theta)\cos(\theta)$$

$$+ \sin(\alpha)\cos(\beta)\sin(\theta)\cos(\theta) + \sin(\alpha)\sin(\beta)\cos(\theta)^2 \ .$$

For the denominator $\sin(\alpha + \theta)\sin(\beta + \theta)$, we first expand one term, and then the other, to get

$$(\cos(\alpha)\sin(\theta) + \sin(\alpha)\cos(\theta))\sin(\beta + \theta) \ ,$$

and then

$$(\cos(\alpha)\sin(\theta) + \sin(\alpha)\cos(\theta)) \cdot (\cos(\beta)\sin(\theta) + \sin(\beta)\cos(\theta)) \ .$$

When multiplied out, this is

$$\cos(\alpha)\cos(\beta)\sin(\theta)^2 + \cos(\alpha)\sin(\beta)\sin(\theta)\cos(\theta)$$

$$+ \sin(\alpha)\cos(\beta)\sin(\theta)\cos(\theta) + \sin(\alpha)\sin(\beta)\cos(\theta)^2 \ ,$$

which is the same as the numerator.

This problem may also be solved using inversion.

Problem 11.13. (This is Problem 2.15 on page 33.) *Prove the following, known as* **Mollweide's Formulae***:*

$$\frac{a+b}{c} = \frac{\cos\left(\frac{A-B}{2}\right)}{\sin\left(\frac{C}{2}\right)}$$

$$\frac{a-b}{c} = \frac{\sin\left(\frac{A-B}{2}\right)}{\cos\left(\frac{C}{2}\right)} \ .$$

Solution. $\quad \dfrac{a+b}{c} = \dfrac{a}{c} + \dfrac{b}{c} = \dfrac{\sin(A)}{\sin(C)} + \dfrac{\sin(B)}{\sin(C)}$

$$= \frac{2\sin\left(\frac{A+B}{2}\right)\cos\left(\frac{A-B}{2}\right)}{2\cos\left(\frac{C}{2}\right)\sin\left(\frac{C}{2}\right)} = \frac{\cos\left(\frac{A-B}{2}\right)}{\sin\left(\frac{C}{2}\right)}$$

$$\frac{a-b}{c} = \frac{a}{c} - \frac{b}{c} = \frac{\sin(A)}{\sin(C)} - \frac{\sin(B)}{\sin(C)}$$

$$= \frac{2\cos\left(\frac{A+B}{2}\right)\sin\left(\frac{A-B}{2}\right)}{2\cos\left(\frac{C}{2}\right)\sin\left(\frac{C}{2}\right)} = \frac{\sin\left(\frac{A-B}{2}\right)}{\cos\left(\frac{C}{2}\right)} \ .$$

Problem 11.14. (This is Problem 2.16 on page 34.) *Prove the following known as the* **Tangent Theorems***:*

$$\frac{a+b}{a-b} = \frac{\tan\left(\frac{A+B}{2}\right)}{\tan\left(\frac{A-B}{2}\right)}$$

$$= \frac{\cot\left(\frac{C}{2}\right)}{\tan\left(\frac{A-B}{2}\right)} \ .$$

Solution. This is easily obtained from the results of the previous problem.

Problem 11.15. (This is Problem 2.13 on page 33.) *Find as many different proofs of the Cosine Rule as you can by searching through books.*

Solution. How many did you find?

Problem 11.16. (This is Problem 2.17 on page 34.) *Show that* $\Delta = 2R^2 \sin(A)\sin(B)\sin(C)$.

Solution. We use the formula just obtained: $\Delta = \frac{abc}{4R}$ together with the **sine rule**:

$$\frac{a}{\sin(A)} = \frac{b}{\sin(B)} = \frac{c}{\sin(C)} = 2R$$

to get this result.

Problem 11.17. (This is Problem 2.18 on page 37.) *Obtain the following formulae:*

(1) $\cos^{-1}(x) + \sin^{-1}(x) = \frac{\pi}{2}$;
(2) $\sin^{-1}(-x) = -\sin^{-1}(x)$;
(3) $\cos^{-1}(-x) = \pi - \cos^{-1}(x)$;
(4) $\sin^{-1}(x) \pm \sin^{-1}(y) = \sin^{-1}\left(x\sqrt{1-y^2} \pm y\sqrt{1-x^2}\right)$
    for $x^2 + y^2 \leq 1$;
(5) $\cos^{-1}(x) + \cos^{-1}(y) = \cos^{-1}\left(xy - \sqrt{(1-x^2)(1-y^2)}\right)$
    for $x^2 + y^2 \geq x^2y^2$;
(6) $\cos^{-1}(x) - \cos^{-1}(y) = -\cos^{-1}\left(xy + \sqrt{(1-x^2)(1-y^2)}\right)$
    for $x \geq y$;
(7) $\cos^{-1}(x) - \cos^{-1}(y) = \cos^{-1}\left(xy + \sqrt{(1-x^2)(1-y^2)}\right)$
    for $x < y$;
(8) $\cos\left(\cos^{-1}(x)\right) = x$;
(9) $\sin\left(\sin^{-1}(x)\right) = x$.

Solution. Most of these can be obtained using triangles. But care is necessary over the ranges.

(1) $\cos^{-1}(x) + \sin^{-1}(x) = \frac{\pi}{2}$;
We start with a diagram where $\sin^{-1}(x) = \theta$:

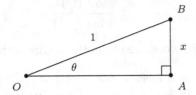

Fig. 11.5   For Problem 11.17 (4) − 1.

We can easily see that $\cos^{-1}(x) = \pi - \theta$.

(2) $\sin^{-1}(-x) = -\sin^{-1}(x)$;
This is easily obtained from the definition of $\sin^{-1}(x)$.

(3) $\cos^{-1}(-x) = \pi - \cos^{-1}(x)$;
This is easily obtained from the definition of $\cos^{-1}(x)$.

(4) $\sin^{-1}(x) \pm \sin^{-1}(y) = \sin^{-1}\left(x\sqrt{1-y^2} \pm y\sqrt{1-x^2}\right)$
for $x^2 + y^2 \leq 1$.
Note that:

$$\theta = \sin^{-1}(x) \longleftrightarrow x = \sin(\theta) \text{ and } -\pi/2 \leq \theta \leq \pi/2,$$
$$\phi = \cos^{-1}(x) \longleftrightarrow x = \cos(\phi) \text{ and } 0 \leq \phi \leq \pi .$$

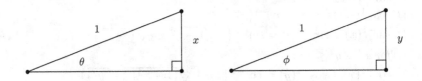

Fig. 11.6   For Problem 11.17 (4) − 2.

Noting that $\sin(\theta \pm \phi) = \sin(\theta)\cos(\phi) \pm \cos(\theta)\sin(\phi)$, we see that

$$\sin(\theta \pm \phi) = x\sqrt{1-y^2} \pm y\sqrt{1-x^2} .$$

To get the restrictions, we consider the following diagram where $\sin^{-1}(x) = \theta$ and $\sin^{-1}(y) = \phi$.

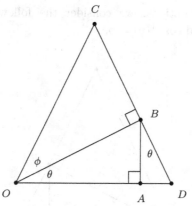

Fig. 11.7 For Problem 11.17 (4) – 3.

Here we let $OB = 1$ and $BC = x$. Thus $OA = \sqrt{1 - x^2}$, $OC = \frac{1}{\sqrt{1-y^2}}$ and $CB = \frac{y}{\sqrt{1-y^2}}$.

Since $\triangle BOA$ is similar to $\triangle DBA$, we have both $BD = \frac{x}{\sqrt{1-x^2}}$ and $AD = \frac{x^2}{\sqrt{1-x^2}}$.

Since $|\theta + \phi| \leq \pi/2$, we must have $CD^2 \leq OC^2 + OD^2$. This is equivalent to:

$$\left(\frac{x}{\sqrt{1 - x^2}} + \frac{y}{\sqrt{1 - y^2}} \right)^2 \leq \left(\frac{1}{\sqrt{1 - y^2}} \right)^2$$
$$+ \left(\sqrt{1 - x^2} + \frac{x^2}{\sqrt{1 - x^2}} \right)^2$$

or

$$\frac{x^2}{1 - x^2} + \frac{y^2}{1 - y^2} + \frac{2xy}{\sqrt{1 - x^2}\sqrt{1 - y^2}} \leq \frac{1}{1 - y^2} + \frac{1}{1 - x^2} \left((1 - x^2) + x^2 \right)^2$$

or

$$xy \leq \sqrt{(1 - x^2)(1 - y^2)}$$

or

$$x^2 + y^2 \leq 1 \ .$$

For the other part, use part 2 of this problem.

(5) $\cos^{-1}(x) + \cos^{-1}(y) = \cos^{-1} \left(xy - \sqrt{(1 - x^2)(1 - y^2)} \right)$
for $x^2 + y^2 \geq x^2 y^2$;

The equality is obtain in a similar way to the start of part 4 of this problem.

To get the restrictions, we consider the following diagram where $\cos^{-1}(x) = \theta$ and $\cos^{-1}(y) = \phi$.

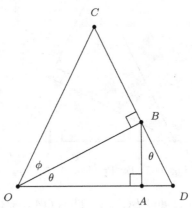

Fig. 11.8   For Problem 11.17 (5).

Here we let $OB = 1$ and $OA = x$. Thus, $AB = \sqrt{1 - x^2}$, $OC = \frac{1}{y}$ and $CB = \frac{\sqrt{1-y^2}}{y}$.

Since $\triangle BOA$ is similar to $\triangle DBA$, we have both $BD = \frac{\sqrt{1-x^2}}{x}$ and $AD = \frac{1}{x} - x$.

We note that $\cos(\theta + \phi) = \frac{OC^2 + OD^2 - DC^2}{2 \cdot OC \cdot OD}$ and it is necessary that this lie in the interval $[-1, 1]$.

This leads to

$$|OC^2 + OD^2 - DC^2| \leq 2|OC| \cdot |OD|$$

or

$$\left| \frac{1}{x^2} + \frac{1}{y^2} - \left(\frac{1}{x^2} - 1 + \frac{1}{y^2} - 1 + \frac{2\sqrt{(1-x^2)(1-y^2)}}{xy} \right) \right| \leq 2 \left| \frac{1}{x} \right| \cdot \left| \frac{1}{y} \right|$$

or

$$\sqrt{(1-x^2)(1-y^2)} \leq 1 \ .$$

(6) $\cos^{-1}(x) - \cos^{-1}(y) = -\cos^{-1}\left(xy + \sqrt{(1-x^2)(1-y^2)} \right)$ for $x \geq y$;

Apply the triangle techniques used in part 4 to get the formula for this part and the next part. The restrictions follow almost immediately.

(7) $\cos^{-1}(x) - \cos^{-1}(y) = \cos^{-1}\left(xy + \sqrt{(1-x^2)(1-y^2)} \right)$ for $x < y$;

(8) $\cos\left(\cos^{-1}(x)\right) = x$;

This and the next part follow immediately from the definitions.

(9) $\sin\left(\sin^{-1}(x)\right) = x$.

Problem 11.18. (This is Problem 2.19 on page 38.) *Which of the following are true, which are false. Justify your answer!*

$$\cos\left(\cos^{-1}(x)\right) = x \qquad\qquad \cos^{-1}\left(\cos(\theta)\right) = \theta \;.$$

$$\sin\left(\sin^{-1}(x)\right) = x \qquad\qquad \sin^{-1}\left(\sin(\theta)\right) = \theta \;.$$

Solution. The two on the left are answered (true) in the previous problem. The two on the right are false as identities: for example, take $\theta = 257\pi$.

Problem 11.19. (This is Problem 2.20 on page 38.) *Show that* $\tan^{-1}(x) = \sin^{-1}\left(\frac{x}{\sqrt{1+x^2}}\right)$.

Solution. We look at the right triangle with "legs" x and 1:

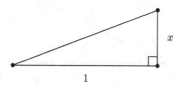

Fig. 11.9   For Problem 11.19.

The hypotenuse is of length $\sqrt{1+x^2}$, giving the formula. We do not have to worry about range restrictions here.

Problem 11.20. (This is Problem 2.21 on page 39.) *Obtain the following formulae:*

(1) $\tan^{-1}(-x) = -\tan^{-1}(x)$;
(2) $\cot^{-1}(-x) = \pi - \cot^{-1}(x)$;
(3) $\tan^{-1}(x) + \tan^{-1}(y) = \tan^{-1}\left(\frac{x+y}{1-xy}\right)$
   for $xy < 1$;
(4) $\tan^{-1}(x) - \tan^{-1}(y) = \tan^{-1}\left(\frac{x-y}{1+xy}\right)$
   for $xy > -1$;
(5) $\cot^{-1}(x) + \cot^{-1}(y) = \cot^{-1}\left(\frac{xy-1}{x+y}\right)$
   for $x + y \neq 0$;
(6) $\cot^{-1}(x) - \cot^{-1}(y) = \cot^{-1}\left(\frac{xy+1}{y-x}\right)$
   for $x \neq y$.

Solution. (1) $\tan^{-1}(-x) = -\tan^{-1}(x)$;

This, and the next part follow directly from the definitions.

(2) $\cot^{-1}(-x) = \pi - \cot^{-1}(x)$;

(3) $\tan^{-1}(x) + \tan^{-1}(y) = \tan^{-1}\left(\frac{x+y}{1-xy}\right)$ for $xy < 1$;

The formula is readily obtained from the usual tangent formulae.

To get the restrictions, we consider the following diagram where $\tan^{-1}(x) = \theta$ and $\tan^{-1}(y) = \phi$.

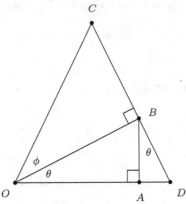

Fig. 11.10   For Problem 11.20 (3).

Here we let $BA = x$ and $OA = 1$. Thus $OB = \sqrt{1+x^2}$, $BC = y\sqrt{1+x^2}$ and $OC = \sqrt{1+x^2}\sqrt{1+y^2}$. Since $\triangle BOA$ is similar to $\triangle DBA$, we have $BD = x\sqrt{1+x^2}$ and $AD = x^2$.

Since $|\theta + \phi| \leq \pi/2$, we must have $CD^2 \leq OC^2 + OD^2$. This is equivalent to:

$$\left(x\sqrt{1+x^2} + y\sqrt{1+x^2}\right)^2 < \left(\sqrt{1+x^2}\sqrt{1+y^2}\right)^2 + (1+x^2)^2$$

or

$$(x+y)^2(1+x^2) < (1+x^2)(1+y^2) + (1+x^2)^2$$

or

$$(x+y)^2 < (1+x^2) + (1+y^2)$$

or

$$xy < 1 \ .$$

(4) $\tan^{-1}(x) - \tan^{-1}(y) = \tan^{-1}\left(\frac{x-y}{1+xy}\right)$

for $xy > -1$;

This is similar to the last part.

(5) $\cot^{-1}(x) + \cot^{-1}(y) = \cot^{-1}\left(\frac{xy-1}{x+y}\right)$
for $x + y \neq 0$;

The required result is equivalent to:

$$(\pi/2 - \tan^{-1}(x)) + (\pi/2 - \tan^{-1}(y)) = \pi/2 - \tan^{-1}\left(\frac{xy-1}{x+y}\right)$$
$$= \pi/2 + \tan^{-1}\left(\frac{1-xy}{x+y}\right) ,$$

or

$$\tan^{-1}(x) + \tan^{-1}(y) = \frac{\pi}{2} - \tan^{-1}\left(\frac{1-xy}{x+y}\right) .$$

We note that $\tan^{-1}(1/z) = \pi/2 - \tan^{-1}(z)$, so that the formula is obtained.

(6) $\cot^{-1}(x) - \cot^{-1}(y) = \cot^{-1}\left(\frac{xy+1}{y-x}\right)$
for $x \neq y$.

This is similar to the last part.

Problem 11.21. (This is Problem 2.22 on page 39.) *Which of the following are true, which are false. Justify your answer!*

$$\tan\left(\tan^{-1}(x)\right) = x \qquad\qquad \tan^{-1}(\tan(\theta)) = \theta .$$

Solution. The first follows from the definition. The second may not be true: for example, try $\theta = 2\pi + \pi/4$.

Problem 11.22. (This is Problem 2.23 on page 40.) *Show that*
$\mathbf{cis}(\theta)\,\mathbf{cis}(\phi)\,\mathbf{cis}(\psi) = \mathbf{cis}(\theta + \phi + \psi)$.
Show that, for any integer n,

(1) $\prod_{k=1}^{n} \mathbf{cis}(\theta_k) = \mathbf{cis}\left(\prod_{k=1}^{n} \theta_k\right)$;
(2) $\left(\mathbf{cis}(\theta)\right)^{n} = \mathbf{cis}(n\theta)$. This is **De Moivre's** Theorem.

Solution. From the text, we have $\mathbf{cis}(\theta)\,\mathbf{cis}(\beta) = \mathbf{cis}(\theta + \beta)$. Now let $\beta = \phi + \psi$, and use the result in the text again to get the result.

(1) This is an extension of the result just obtained, and should be proved by induction.
(2) Put $\theta_k = \theta$ in part 1 above to get this.

Problem 11.23. (This is Problem 2.24 on page 40.) *Use De Moivre's theorem to prove that*

$$\left(\operatorname{cis}(\theta)\right)^{1/n} = \operatorname{cis}\left(\frac{\theta}{n}\right)$$

for every natural number n.
Deduce De Moivre's theorem for every rational number n.

Solution. From the previous problem, we have (with ϕ replacing θ),

$$\left(\operatorname{cis}(\phi)\right)^{n} = \operatorname{cis}(n\phi) \ .$$

In this, let $n\phi = \theta$, to give

$$\left(\operatorname{cis}\left(\frac{\theta}{n}\right)\right)^{n} = \operatorname{cis}(\theta) ,$$

and the result follows.
By combining this result (with m replacing n) and De Moivre's Theorem (with p replacing n), we get

$$\left(\operatorname{cis}(\theta)\right)^{p/m} = \operatorname{cis}\left(\frac{p}{m}\theta\right)$$

or

$$\left(\operatorname{cis}(\theta)\right)^{n} = \operatorname{cis}(n\theta)$$

for every positive rational n.
It is also easy to show that

$$(\operatorname{cis}(\theta))^{-1} = \frac{1}{\cos(\theta) + i\,\sin(\theta)} = \cos(\theta) - i\,\sin(\theta)$$
$$= \cos(-\theta) + i\,\sin(-\theta) = \operatorname{cis}(-t) \ .$$

Combining this with the previous result on positive rationals gives De Moivre's Theorem for all rationals.

Problem 11.24. (This is Problem 2.25 on page 41.) *Prove the following formulae:*

(1) $\sin(3\theta) = 3\sin(\theta) - 4\sin^3(\theta)$;
(2) $\cos(3\theta) = 4\cos^3(\theta) - 3\cos(\theta)$;
(3) $\sin(4\theta) = 8\sin(\theta)\cos^3(\theta) - 4\sin(\theta)\cos(\theta)$;
(4) $\cos(5\theta) = 16\cos^5(\theta) - 20\cos^3(\theta) + 5\cos(\theta)$;
(5) $\tan(3\theta) = \frac{3\tan(\theta) - \tan^3(\theta)}{1 - 3\tan^2(\theta)}$;
(6) $\cot(4\theta) = \frac{\cot^4(\theta) - 6\cot^2(\theta) + 1}{4\cot^3(\theta) - 4\cot(\theta)}$.

Solution. (1) We have from the text

$$\sin(3\theta) = 3\cos^2(\theta)\sin(\theta) - \sin^3(\theta)$$
$$= 3\left(1 - \sin^2(\theta)\right)\sin(\theta) - \sin^3(\theta)$$
$$= 3\sin(\theta) - 4\sin^3(\theta) \ .$$

(2) We have from the text

$$\cos(3\theta) = \cos^3(\theta) - 3\cos(\theta)\sin^2(\theta)$$
$$= \cos^3(\theta) - 3\cos(\theta)\left(1 - \cos^2(\theta)\right)$$
$$= 4\cos^3(\theta) - 3\cos(\theta) \ .$$

(3) We start with De Moivre's Theorem: $\mathbf{cis}(4\theta) = \left(\mathbf{cis}(\theta)\right)^4$. This leads to

$$\cos(4\theta) + i\,\sin(4\theta) = \left(\cos^4(\theta) - 6\cos^2(\theta)\sin^2(\theta) + \sin^4(\theta)\right)$$
$$+i\left(4\cos^3(\theta)\sin(\theta) - 4\cos(\theta)\sin^3(\theta)\right) \ .$$

From which we get

$$\sin(4\theta) = 4\cos^3(\theta)\sin(\theta) - 4\cos(\theta)\sin^3(\theta) \ .$$

We now substitute for $\sin^2(\theta)$ to get

$$\sin(4\theta) = 8\sin(\theta)\cos^3(\theta) - 4\sin(\theta)\cos(\theta) \ .$$

(4) We start with De Moivre's Theorem: $\mathbf{cis}(5\theta) = \left(\mathbf{cis}(\theta)\right)^5$. This leads to

$$\cos(5\theta) + i\,\sin(5\theta) = \left(\cos^5(\theta) - 10\cos^3(\theta)\sin^2(\theta) + 5\cos(\theta)\sin^4(\theta)\right)$$
$$+i\left(5\cos^4(\theta)\sin(\theta) - 10\cos^2(\theta)\sin^3(\theta) + \sin^5(\theta)\right) \ .$$

From which we get

$$\cos(5\theta) = \cos^5(\theta) - 10\cos^3(\theta)\sin^2(\theta) + 5\cos(\theta)\sin^4(\theta) \ .$$

We now substitute for $\sin^2(\theta)$ to get

$$\cos(5\theta) = 16\cos^5(\theta) - 20\cos^3(\theta) + 5\cos(\theta) \ .$$

(5) Using the results for $\sin(3\theta)$ and $\cos(3\theta)$ in earlier parts of this problem, we get

$$\tan(3\theta) = \frac{3\sin(\theta) - 4\sin^3(\theta)}{4\cos^3(\theta) - 3\cos(\theta)} .$$

We divide top and bottom by $\cos^3(\theta)$ to get

$$\tan(3\theta) = \frac{3\tan(\theta) - \tan^3(\theta)}{1 - 3\tan^2(\theta)} .$$

(6) In part 3 above, we see how to get the expressions for $\cos(4\theta)$ and $\sin(4\theta)$. As in part 5, we get

$$\cot(4\theta) = \frac{\cot^4(\theta) - 6\cot^2(\theta) + 1}{4\cot^3(\theta) - 4\cot(\theta)} .$$

Problem 11.25. (This is Problem 2.26 on page 41.) *Find general formulae for* $\sin(n\theta)$ *and* $\cos(n\theta)$. (*You will need to make use of binomial coefficients.*)

Solution. The binomial theorem states:

$$(x + y)^n = \sum_{k=0}^{n} \binom{n}{k} x^k y^{n-k} .$$

With $x = \cos(\theta)$ and $y = i\sin(\theta)$, we get

$$\cos(n\theta) + i\sin(n\theta) = \sum_{k=0}^{n} \binom{n}{k} \cos^k(\theta) i^{n-k} \sin^{n-k}(\theta) .$$

Since i^m ($m = 0, 1, 2, \ldots$) runs through the sequence $1, i, -1, -i, 1, i, -1, -i, \ldots$, we have

$$\big(\cos(\theta) + i\sin(\theta)\big)^n$$
$$= \left(\cos^n(\theta) - \binom{n}{2}\cos^{n-2}(\theta)\sin^2(\theta) + \binom{n}{4}\cos^{n-4}(\theta)\sin^4(\theta) + \ldots \right)$$
$$+ i\left(\binom{n}{1}\cos^{n-1}(\theta)\sin(\theta) - \binom{n}{3}\cos^{n-3}(\theta)\sin^3(\theta) + \ldots \right) .$$

These give the two general formulae. Note that the form of the last term depends on whether n is odd or even.

Problem 11.26. (This is Problem 2.27 on page 41.) *Prove the following formulae:*

(1) $\sin^3(\theta) = \frac{1}{4}\left(3\sin(\theta) - \sin(3\theta) \right)$;

(2) $\cos^3(\theta) = \frac{1}{4}\left(3\cos(\theta) + \cos(3\theta) \right)$;

(3) $\sin^4(\theta) = \frac{1}{8}\left(\cos(4\theta) - 4\cos(2\theta) + 3 \right)$;

(4) $\cos^5(\theta) = \frac{1}{16}\left(10\cos(\theta) + 5\cos(3\theta) + \cos(5\theta) \right)$.

Solution. (1) Since $\sin(3\theta) = 3\sin(\theta) - 4\sin^3(\theta)$,

we have $\sin^3(\theta) = \frac{1}{4}\left(3\sin(\theta) - \sin(3\theta) \right)$;

(2) Since $\cos(3\theta) = 4\cos^3(\theta) - 3\cos(\theta)$, we have

$\cos^3(\theta) = \frac{1}{4}\left(3\cos(\theta) + \cos(3\theta) \right)$;

(3) Since

$$\begin{aligned}
\cos(4\theta) &= \cos^4(\theta) - 6\cos^2(\theta)\sin^2(\theta) + \sin^4(\theta)\\
&= \left(1 - \sin^2(\theta)\right)^2 - 6\sin^2(\theta)\left(1 - \sin^2(\theta)\right) + \sin^4(\theta)\\
&= 1 - 8\sin^2(\theta) + 8\sin^4(\theta),
\end{aligned}$$

we have

$$\begin{aligned}
\sin^4(\theta) &= \frac{1}{8}\left(\cos(4\theta) - 1 \right) + \sin^2(\theta)\\
&= \frac{1}{8}\left(\cos(4\theta) - 1 \right) + \frac{1}{2}\left(1 - \cos(2\theta) \right)\\
&= \frac{1}{8}\left(\cos(4\theta) - 4\cos(2\theta) + 3 \right) \quad .
\end{aligned}$$

(4) Since

$$\begin{aligned}
\cos(5\theta) &= \cos^5(\theta) - 10\cos^3(\theta)\sin^2(\theta) + 5\cos(\theta)\sin^4(\theta)\\
&= \cos^5(\theta) - 10\cos^3(\theta)\left(1 - \cos^2(t)\right) + 5\cos(\theta)\left(1 - \cos^2(\theta)\right)^2\\
&= 16\cos^5(\theta) - 20\cos^3(\theta) + 5\cos(\theta)
\end{aligned}$$

and

$$\cos^3(\theta) = \frac{1}{4}\left(3\cos(\theta) + \cos(3\theta) \right)$$

we obtain that

$$\cos^5(\theta) = \frac{1}{16}\left(10\cos(\theta) + 5\cos(3\theta) + \cos(5\theta)\right) .$$

Problem 11.27. (This is Problem 2.28 on page 42.) *Find the parametric equations for the ellipse in standard position*

$$\frac{x^2}{a^2} + \frac{y^2}{b^2} = 1 .$$

Solution. Let $x = a\cos(\theta)$ and $y = b\sin(\theta)$.

Problem 11.28. (This is Problem 2.29 on page 43.) *Prove that, for $0 < \theta_k < \pi$,*

$$\left|\sin\left(\sum_{k=1}^{\mu}\theta_k\right)\right| < \sum_{k=1}^{\mu}\sin(\theta_k) .$$

Solution. Consider first the case when $n = 2$.
We note that

$$\sin(x) + \sin(y) = 2\sin\left(\frac{x+y}{2}\right)\cos\left(\frac{x-y}{2}\right)$$

and that

$$\sin(x+y) = 2\sin\left(\frac{x+y}{2}\right)\cos\left(\frac{x+y}{2}\right) .$$

Thus, we must show that, for $0 < x, y < \pi$,

$$\sin(x+y) - |\sin(x) + \sin(y)|$$
$$= 2\sin\left(\frac{x+y}{2}\right)\left(\cos\left(\frac{x-y}{2}\right) - \left|\cos\left(\frac{x+y}{2}\right)\right|\right) > 0 .$$

Now, $0 < x + y < 2\pi$, so that $\sin\left(\frac{x+y}{2}\right) > 0$. This means that we must show that, for $0 < x, y < \pi$,

$$\cos\left(\frac{x-y}{2}\right) - \left|\cos\left(\frac{x+y}{2}\right)\right| > 0 .$$

Now, there are two case to consider:

$$\cos\left(\frac{x-y}{2}\right) - \cos\left(\frac{x+y}{2}\right) = 2\sin(x/2) * \sin(y/2) > 0 ,$$

since both $\sin(x/2) > 0$ and $\sin(y/2) > 0$ for $0 < x, y < \pi$, and

$$\cos\left(\frac{x-y}{2}\right) + \cos\left(\frac{x+y}{2}\right) = 2\cos(x/2)\cos(y/2) > ,$$

since both $\cos(x/2) > 0$ and $\cos(y/2) > 0$ for $0 < x, y < \pi$.
This gives the essence of a proof by induction for the general case.

Problem 11.29. (This is Problem 2.30 on page 43.) *If $S_c = \sum\limits_{k=\lambda+1}^{\mu} \cos(k\theta)$ and*

$S_s = \sum\limits_{k=\lambda+1}^{\mu} \sin(k\theta)$, *show that, for* $0 < \theta < 2\pi$,

$$\sqrt{S_c^2 + S_s^2} \leq \frac{1}{\sin(\theta/2)} \ .$$

Solution. In the text, we have the two formulae:

$$S_c = \sum_{k=\lambda+1}^{\mu} \cos(k\theta) = \left(\frac{\cos\left\{(\mu+\lambda+1)\theta/2\right\} \sin\left\{(\mu-\lambda)\theta/2\right\}}{\sin(\theta/2)} \right),$$

$$S_s = \sum_{k=\lambda+1}^{\mu} \sin(k\theta) = \left(\frac{\sin\left\{(\mu+\lambda+1)\theta/2\right\} \sin\left\{(\mu-\lambda)\theta/2\right\}}{\sin(\theta/2)} \right).$$

Thus,

$$\sin^2(\theta/2) \left(S_c^2 + S_s^2 \right)$$
$$= \sin^2\left((\mu-\lambda)\theta/2\right) \left(\sin^2\left((\mu+\lambda+1)\theta/2\right) + \cos^2\left((\mu+\lambda+1)\theta/2\right) \right)$$
$$= \sin^2\left((\mu-\lambda)\theta/2\right) \leq 1 \ .$$

Problem 11.30. (This is Problem 2.31 on page 44.) *Show the following properties for all $x \in R$:*

(1) $\sinh(x) \in (-\infty, \infty)$;
(2) $\cosh(x) \in [1, \infty)$;
(3) $\tanh(x) \in (-1, 1)$.

Solution. (1) $\sinh(x) = \frac{e^x - e^{-x}}{2} \to \infty$ as $x \to \infty$,
and $\sinh(x) = \frac{e^x - e^{-x}}{2} \to -\infty$ as $x \to -\infty$.
Also, $\sinh(x)$ is continuous on its domain.
(2) $\cosh(x) = \frac{e^x + e^{-x}}{2} \to \infty$ as $|x| \to \infty$. We also note that

$$2\cosh(x) - 2 = e^x - 2 + e^{-x} = \left(\sqrt{e^x} - \sqrt{e^{-x}} \right)^2 \geq 0 \ ,$$

which shows that $\cosh(x) \geq 1$.
(3) $\tanh(x) = \frac{e^x - e^{-x}}{e^x + e^{-x}} \to 1$ as $x \to \infty$,
and $\tanh(x) = \frac{e^x - e^{-x}}{e^x + e^{-x}} \to -1$ as $x \to -\infty$.
Also, $\tanh(x)$ is continuous on its domain.

Problem 11.31. (This is Problem 2.32 on page 45.) *Show that*

(1) $\coth(x) \in (1, \infty)$ *for* $x \in (0, \infty)$;
(2) $\coth(x) \in (-\infty, -1)$ *for* $x \in (-\infty, 0)$;
(3) $\displaystyle\lim_{x \to \infty} \tanh(x) = \lim_{x \to \infty} \coth(x) = 1$;
(4) $\displaystyle\lim_{x \to -\infty} \tanh(x) = \lim_{x \to -\infty} \coth(x) = -1$.

Solution. These are obtained from part 3 of the previous question.

Problem 11.32. (This is Problem 2.33 on page 45.) *Show that*

(1) $\sinh(2x) = 2\sinh(x)\cosh(x)$;
(2) $\cosh(2x) = \cosh^2(x) + \sinh^2(x)$;
(3) $\tanh(2x) = \frac{2\tanh(x)}{1 + \tanh^2(x)}$.

Find two other expressions for $\cosh(2x)$.

Solution.

(1)

$$2\sinh(x)\cosh(x) = 2\frac{e^x - e^{-x}}{2}\frac{e^x + e^{-x}}{2}$$
$$= \frac{e^{2x} - e^{-2x}}{2} = \sinh(2x)$$

(2)
$$\cosh^2(x) + \sinh^2(x) = \left(\frac{e^x + e^{-x}}{2}\right)^2 + \left(\frac{e^x - e^{-x}}{2}\right)^2$$
$$= \frac{e^{2x} + 2 + e^{-2x}}{4} + \frac{e^{2x} - 2 + e^{-2x}}{4}$$
$$= \frac{e^{2x} + e^{-2x}}{2} = \cosh(2x)$$

(3)
$$\tanh(2x) = \frac{\sinh(2x)}{\cosh(2x)} = \frac{2\sinh(x)\cosh(x)}{\cosh^2(x) + \sinh^2(x)} = \frac{2\tanh(x)}{1 + \tanh^2(x)} \quad .$$

the last line being obtained by dividing top and bottom by $\cosh^2(x)$.

Since $\cosh(2x) = \cosh^2(x) + \sinh^2(x)$ and $1 = \cosh^2(x) - \sinh^2(x)$, we have

$$\cosh(2x) = 2\cosh^2(x) - 1 = 2\sinh^2(x) + 1 \quad .$$

Problem 11.33. (This is Problem 2.34 on page 45.) *Translate all the trigonometric identities in exercises (11.24) and (11.26) above to hyperbolic identities.*

Solution. Problems similar to 11.24.
(1) $\sinh(3\theta) = 3\sinh(\theta) + 4\sinh^3(\theta)$;
(2) $\cosh(3\theta) = 4\cosh^3(\theta) - 3\cosh(\theta)$;
(3) $\sinh(4\theta) = 8\sinh(\theta)\cosh^3(\theta) - 4\sinh(\theta)\cosh(\theta)$;
(4) $\cosh(5\theta) = 16\cosh^5(\theta) - 20\cosh^3(\theta) + 5\cosh(\theta)$;
(5) $\tanh(3\theta) = \frac{3\tanh(\theta)+\tanh^3(\theta)}{1+3\tanh^2(\theta)}$;
(6) $\coth(4\theta) = \frac{\coth^4(\theta)+6\coth^2(\theta)+1}{4\coth^3(\theta)+4\coth(\theta)}$.

Problems similar to 11.26.
(1) $\sinh^3(\theta) = \frac{1}{4}\left(-3\sinh(\theta) + \sinh(3\theta) \right)$;

(2) $\cosh^3(\theta) = \frac{1}{4}\left(3\cosh(\theta) + \cosh(3\theta) \right)$;

(3) $\sinh^4(\theta) = \frac{1}{8}\left(\cosh(4\theta) - 4\cosh(2\theta) + 3 \right)$;

(4) $\cosh^5(\theta) = \frac{1}{16}\left(10\cosh(\theta) + 5\cosh(3\theta) + \cosh(5\theta) \right)$.

Problem 11.34. (This is Problem 2.35 on page 46.) *Obtain the following formulae:*

(1) $\sinh^{-1}(x) = \cosh^{-1}\left(\sqrt{x^2+1}\right)$ *for $x \geq 0$* ;
(2) $\sinh^{-1}(x) = -\cosh^{-1}\left(\sqrt{x^2+1}\right)$ *for $x < 0$* ;
(3) $\sinh^{-1}(x) = \tanh^{-1}\left(\frac{x}{\sqrt{x^2+1}}\right)$;
(4) $\sinh^{-1}(x) = \coth^{-1}\left(\frac{\sqrt{x^2+1}}{x}\right)$;
(5) $\cosh^{-1}(x) = \left|\sinh^{-1}\left(\sqrt{x^2-1}\right)\right|$;
(6) $\cosh^{-1}(x) = \left|\tanh^{-1}\left(\frac{\sqrt{x^2-1}}{x}\right)\right|$;
(7) $\cosh^{-1}(x) = \left|\coth^{-1}\left(\frac{x}{\sqrt{x^2-1}}\right)\right|$;
(8) $\tanh^{-1}(x) = \sinh^{-1}\left(\frac{x}{\sqrt{1-x^2}}\right)$;
(9) $\tanh^{-1}(x) = \cosh^{-1}\left(\frac{1}{\sqrt{1-x^2}}\right)$ *for $x > 0$* ;
(10) $\tanh^{-1}(x) = -\cosh^{-1}\left(\frac{1}{\sqrt{1-x^2}}\right)$ *for $x < 0$* .

Solution. Note that we cannot work with triangles!

(1) $y = \sinh^{-1}(x)$ if and only if $x = \sinh(y)$.

Thus, $\cosh(y) = \sqrt{x^1 + 1}$, giving $y = \pm \cosh^{-1}\left(\sqrt{x^2 + 1}\right)$. The sign depends on whether x is positive or negative. In this part $x \geq 0$, so that we have $y = \cosh^{-1}\left(\sqrt{x^2 + 1}\right)$.

(2) This is, of course like the previous part, except that in this part $x < 0$, giving $y = -\cosh^{-1}\left(\sqrt{x^2 + 1}\right)$.

(3) We note that $\tanh(y) = \frac{\sinh(y)}{\cosh(y)} = \frac{\sinh(y)}{\sqrt{\sinh^2(y)+1}}$.

Setting $x = \sinh(y)$ gives the result.

(4) Use the definition of $\coth^{-1}(x)$ and the previous part.

The remainder of this problem is just other perspectives on these parts.

Problem 11.35. (This is Problem 2.36 on page 47.) *Obtain the following formulae:*

(1) $\sinh^{-1}(x) \pm \sinh^{-1}(y) = \sinh^{-1}\left(x\sqrt{1+y^2} \pm y\sqrt{1+x^2}\right)$;

(2) $\cosh^{-1}(x) \pm \cosh^{-1}(y) = \cosh^{-1}\left(xy \pm \sqrt{(1+x^2)(1+y^2)}\right)$;

(3) $\tanh^{-1}(x) \pm \tanh^{-1}(y) = \tanh^{-1}\left(\frac{x \pm y}{1 \pm xy}\right)$.

Solution. (1) Let $\alpha = \sinh(x)$ and $\beta = \sinh(y)$.

Then $\sinh(\alpha + \beta) = \sinh(\alpha)\cosh(y) + \cosh(x)\sinh(y)$

We note that

$$\cosh(\alpha) = \sqrt{1 + \sinh^2(\alpha)} \quad \text{and} \quad \cosh(\beta) = \sqrt{1 + \sinh^2(\beta)} .$$

This leads to the result.

All the others parts are done similarly.

Problem 11.36. (This is Problem 2.37 on page 47.) *Obtain the following formulae:*

(1) $\sinh^{-1}(i\theta) = i\sin^{-1}(\theta)$;

(2) $\cosh^{-1}(i\theta) = i\cos^{-1}(\theta)$;

(3) $\tanh^{-1}(i\theta) = i\tan^{-1}(\theta)$.

Solution. These follow from earlier results in this book.

Problem 11.37. (This is Problem 2.38 on page 48.) *If you have the background Calculus, derive the four formulae:*

$$\sinh^{-1}(x) = \log\left(x + \sqrt{x^2 + 1}\right) \; ;$$

$$\cosh^{-1}(x) = \log\left(x + \sqrt{x^2 - 1}\right) \; ;$$

$$\tanh^{-1}(x) = \frac{1}{2}\log\left(\frac{1+x}{1-x}\right) \qquad for \qquad |x| < 1 \; ;$$

$$\coth^{-1}(x) = \frac{1}{2}\log\left(\frac{x+1}{x-1}\right) \qquad for \qquad |x| > 1 \; .$$

Solution. (1) Let $y = \sinh^{-1}(x)$, so that $x = \sinh(y)$. Thus

$$\frac{dx}{dy} = \cosh(y) = \sqrt{1 + \sinh^2(y)},$$

giving

$$\frac{dy}{dx} = \frac{1}{\sqrt{1 + \sinh^2(y)}} = \frac{1}{\sqrt{1 + x^2}} \; per \; .$$

Also, if $z = \log\left(x + \sqrt{1 + x^2}\right)$, we have

$$\frac{dz}{dx} = \frac{1}{x + \sqrt{1 + x^2}} \cdot \left(1 + \frac{2x}{2\sqrt{1 + x^2}}\right) = \frac{1}{x + \sqrt{1 + x^2}} \cdot \frac{x + \sqrt{1 + x^2}}{\sqrt{1 + x^2}}$$

$$= \frac{1}{\sqrt{1 + x^2}} \; .$$

Thus, $\sinh^{-1}(x)$ and $\log\left(x + \sqrt{1 + x^2}\right)$ differ by at most a constant. Since $\sinh^{-1}(0) = 0$ and $\log\left(0 + \sqrt{1 + 0^2}\right) = \log(1) = 0$, the two expressions are equal.

(2) The other parts are done in a similar manner.

Problem 11.38. (This is Problem 3.1 on page 51.) *Investigate* **degenerate** *cases of Menelaus' Theorem.*
For example, what is the case if l passes through a vertex?
Can we interpret the theorem if l is parallel to a side?

Solution. We shall consider first N as a fixed point and allow the points M and L to move along their sides towards B.

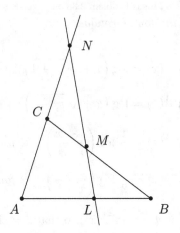

This gives:

$$-1 = \frac{CN}{NA} \cdot \left(\lim_{M \to B} \frac{BM}{MC} \right) \cdot \left(\lim_{L \to B} \frac{AL}{LB} \right) = \frac{CN}{NA} \cdot \frac{AB}{BC} \cdot \left(\lim_{L,M \to B} \frac{BM}{LB} \right) .$$

Since A, B, C and N are fixed points, the value of the limit must be such as to maintain the equality.

However, N was, in a sense, an arbitrary point. Thus, there is in fact, no really meaningful interpretation.

If LM were to be parallel to AC, the N would be "at infinity". We would have

$$-1 = \frac{AL}{LB} \cdot \frac{BM}{MC} \cdot \left(\lim_{N \to \infty} \frac{CN}{NA} \right) .$$

But when $LM \| AC$, we have similar triangles $\triangle ABC$ and $\triangle LMB$ giving $\frac{AL}{LB} = \frac{CM}{MB}$, so that

$$\lim_{N \to \infty} \frac{CN}{NA} = -1 .$$

Problem 11.39. (This is Problem 3.3 on page 53.) *Prove the converse of Ceva's Theorem: that is, if point D, E and F on sides BC, CA and AB, respectively, of triangle $\triangle ABC$ satisfy*

$$\frac{AF}{FB} \cdot \frac{BD}{DC} \cdot \frac{CE}{EA} = 1,$$

then the lines AD, BE and CF are concurrent.

Solution. Suppose that AD and BE meet at P, and suppose that CP meets AB at F'.

The Ceva's theorem gives that

$$\frac{AF'}{F'B} \cdot \frac{BD}{DC} \cdot \frac{CE}{EA} = 1 .$$

But we are given that

$$\frac{AF}{FB} \cdot \frac{BD}{DC} \cdot \frac{CE}{EA} = 1 .$$

Thus,

$$\frac{AF'}{F'B} = \frac{AF}{FB}$$

from which we see that F and F' must be the same point.

Problem 11.40. *We have L, M and N as the mid-points of BC, CA and AB, respectively.*

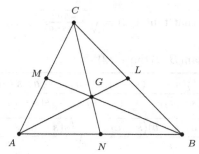

In the above diagram, show that

(1) $ML \parallel AB$,
(2) $ML = \frac{AB}{2}$.
   Find similar results for the other sides of the triangle.

Solution.

(1) It is easy to see that $\triangle CAB$ and $\triangle CML$ are similar
   ($CM = \frac{CA}{2}$, $CL = \frac{CB}{2}$ and $\angle ACB = \angle MCL$).
   It now follows that $\angle LMC = \angle BAC$, so that $ML \parallel AB$,
(2) And it follows that $ML = \frac{AB}{2}$.
   We also have:
   $LN \parallel CA$, $NM \parallel BC$, $LN = \frac{CA}{2}$ and $NM = \frac{BC}{2}$.

Problem 11.41. (This is Problem 4.1 on page 80.) *Prove that*
$$s = 4R\cos(A/2)\cos(B/2)\cos(C/2) \ .$$

Solution. We have that $\cos(A/2) = \sqrt{\frac{s(s-a)}{bc}}, \dots$
Thus,

$$\cos(A/2)\cos(B/2)\cos(C/2)$$
$$= \sqrt{\frac{s(s-a)}{bc}\frac{s(s-b)}{ca}\frac{s(s-c)}{ab}}$$
$$= \sqrt{\frac{s^2 \times s(s-a)(s-b)(s-c)}{a^2b^2c^2}}$$
$$= \sqrt{\frac{s^2\Delta^2}{a^2b^2c^2}} = \frac{s\Delta}{abc} = \frac{s}{4R} \ .$$

Problem 11.42. (This is Problem 4.2 on page 80.) *Prove that*
$$\Delta = s^2 \tan(A/2)\tan(B/2)\tan(C/2) \ .$$

Solution. We have that $\tan(A/2) = \sqrt{\frac{(s-b)(s-c)}{s(s-a)}}, \dots$
Thus,

$$\tan(A/2)\tan(B/2)\tan(C/2)$$
$$= \sqrt{\frac{(s-b)(s-c)}{s(s-a)}\frac{(s-c)(s-a)}{s(s-b)}\frac{(s-a)(s-b)}{s(s-c)}}$$
$$= \sqrt{\frac{(s-a)(s-b)(s-c)}{s^3}} = \sqrt{\frac{s(s-a)(s-b)(s-c)}{s^4}}$$
$$= \sqrt{\frac{\Delta^2}{s^4}} = \frac{\Delta}{s^2} \ .$$

Problem 11.43. (This is Problem 4.3 on page 81.) *Prove that*
$$r = s\tan(A/2)\tan(B/2)\tan(C/2) \ .$$

Solution. We have that $\tan(A/2) = \frac{r}{s-a}, \dots$
Thus,

$$\tan(A/2)\tan(B/2)\tan(C/2)$$
$$= \left(\frac{r}{s-a}\right)\left(\frac{r}{s-b}\right)\left(\frac{r}{s-c}\right)$$
$$= \frac{r^3}{(s-a)(s-b)(s-c)} = \frac{r^3 s}{s(s-a)(s-b)(s-c)}$$
$$= \frac{r^3 s}{\Delta^2} = \frac{r^3 s}{r^2 s^2} = \frac{r}{s} \ .$$

Problem 11.44. (This is Problem 4.4 on page 81.) *Prove that*
$$r = 4R\sin(A/2)\sin(B/2)\sin(C/2) \ .$$

Solution. We have $\sin(A/2) = \sqrt{\frac{(s-b)(s-c)}{bc}}, \dots$
Thus,

$$\sin(A/2)\sin(B/2)\sin(C/2)$$

$$= \sqrt{\frac{(s-b)(s-c)}{bc}}\sqrt{\frac{(s-c)(s-a)}{ca}}\sqrt{\frac{(s-a)(s-b)}{ab}}$$

$$= \sqrt{\frac{((s-a)(s-b)(s-c))^2}{(abc)^2}} = \frac{(s-a)(s-b)(s-c)}{abc}$$

$$= \frac{s(s-a)(s-b)(s-c)}{sabc} = \frac{\Delta^2}{sabc}$$

$$= \frac{r^2 s^2}{sabc} = \frac{r^2 s}{abc} = \frac{r^2 s}{4Rrs} = \frac{r}{4R} \ .$$

Problem 11.45. (This is Problem 4.5 on page 81.) *Prove that*
$$\Delta = r^2 \cot(A/2)\cot(B/2)\cot(C/2) \ .$$

Solution. From problem 11.43, we have
$$r = s\tan(A/2)\tan(B/2)\tan(C/2) \ .$$
Thus, $\cot(A/2)\cot(B/2)\cot(C/2) = \frac{s}{r} = \frac{rs}{r^2} = \frac{\Delta}{r^2}$.

Problem 11.46.
In the given figure (Fig. 11.11), prove that $BR' = z$.

Solution. Note that $AP' \perp P'I_A$. We also note the following:
$I_A R' = I_A A' = I_A P'$ (radii of the escribed circle)

$\triangle I_A R'A$ is congruent to $\triangle I_A P'A$ (right triangles,
common hypotenuse,
equal leg), giving:

$AR' = AP$
Denote $CP' = CA'$ by η, and $BR' = BA'$ by ζ. Note that $\zeta + \eta = BC = z + y$, $AR' = c + \zeta$ and $AP' = b + \eta$.
Solving these equations for ζ and η gives:

$$BR' = \zeta = \frac{a+b-c}{2} = \frac{a+b+c-2c}{2} = \frac{2s-2c}{2} = s-c = z \ ,$$

and, as a bonus,

$$CP' = \eta = \frac{a-b+c}{2} = \frac{a+b+c-2b}{2} = \frac{2s-2b}{2} = s-b = y \ .$$

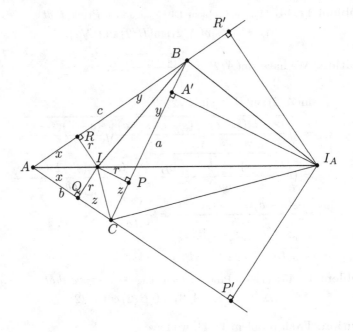

Fig. 11.11

Problem 11.47. (This is Problem 4.7 on page 83.)
Using the same diagram as in the previous problem, prove that

$$
\begin{aligned}
r_A &= s\tan(A/2) = (x+y+z)\tan(A/2) \;, \\
&= (s-c)\cot(B/2) = z\cot(B/2) \;, \\
&= (s-b)\cot(C/2) = y\cot(C/2) \;, \\
&= \frac{a\cos(B/2)\cos(C/2)}{\cos(A/2)} \;, \\
&= \Delta/(s-a) = \Delta/x \;.
\end{aligned}
$$

Solution.
In the diagram (see Fig. 11.11), we see that $AR' = AP' = c + z = s$. Thus,

$$
\begin{aligned}
r_A = I_A R' &= AR'\tan(\angle R'AI_A) \\
&= s\tan(A/2) = (x+y+s)\tan(a/2) \;\textbf{[first result]}
\end{aligned}
$$

$$= BR' \tan(\angle R'BI_A) = z \tan\left(\frac{\angle R'BA'}{2}\right)$$
$$= z \tan\left(\frac{\pi - \angle ABC}{2}\right) = z \tan(\pi/2 - B/2)$$
$$= z \cot(B/2) = (s-c)\cot(B/2) \qquad \text{[second result]}$$

$$= \text{(using a similar argument)}$$
$$= y \cot(C/2) = (s-b)\cot(C/2) \qquad \text{[third result]}$$

$$= AR' \times \frac{IR}{RA} \quad \text{(from similar triangles)}$$
$$= s\frac{r}{x} = \frac{rs}{x} = \frac{\Delta}{x} = \frac{\Delta}{s-a} \qquad \text{[fifth result]}$$

$$= \sqrt{\frac{s(s-b)(s-c)}{s-a}} = a\sqrt{\frac{s(s-b)}{ac}\frac{s(s-c)}{ab}\frac{bc}{s(s-a)}}$$
$$= \frac{a\cos(B/2)\cos(C/2)}{\cos(A/2)} \qquad \text{[fourth result]}$$

Problem 11.48. (This is Problem 4.8 on page 83.) *Suppose that R is the circumradius of triangle ABC and r_A, r_B and r_C are the radii of the escribed circles.*

(1) $r_A r_B + r_B r_C + r_C r_A = s^2$,
(2) $\triangle ABC = \frac{r_A r_B r_C}{s}$.

If the distances between the centres of the escribed circles are α, β and γ, and $\sigma = \frac{\alpha+\beta+\gamma}{2}$, show that

$$8R = \frac{\alpha\beta\gamma}{\sqrt{\sigma(\sigma-\alpha)(\sigma-\beta)(\sigma-\gamma)}} .$$

Solution. (1) From Problem 11.47, we have

$$r_A = \frac{\Delta}{x} \qquad r_B = \frac{\Delta}{y} \qquad r_C = \frac{\Delta}{z} ,$$

so that

$$r_A r_B + r_B r_C + r_C r_A = \Delta^2\left(\frac{1}{xy} + \frac{1}{yz} + \frac{1}{zx}\right)$$
$$= \frac{\Delta^2(x+y+z)}{xyz} = \frac{\Delta^2 s}{\frac{\Delta^2}{s}} = s^2 .$$

(2) From

$$r_A r_B r_C = \frac{\Delta^3}{(xyz)^2} = \frac{\Delta^3}{\frac{\Delta^2}{s}} = \Delta s ,$$

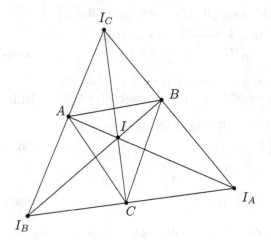

Fig. 11.12

the second result follows.

For the second part, we need a new diagram:

From the previous diagram, we see that $\angle CI_AP' = \pi/2 - \angle I_ACP' = C/2$.

Thus, $I_AC = \frac{y}{\sin(C/2)}$.

This leads to:

$$I_AC = \frac{y}{\sin(C/2)} \qquad I_BC = \frac{x}{\sin(C/2)}$$

$$I_BA = \frac{z}{\sin(A/2)} \qquad I_CA = \frac{y}{\sin(A/2)}$$

$$I_CB = \frac{x}{\sin(B/2)} \qquad I_AB = \frac{z}{\sin(B/2)}$$

$$\alpha = I_BI_C = \frac{y+z}{\sin(A/2)} = \frac{a}{\sin(A/2)}$$

$$\beta = I_CI_A = \frac{z+x}{\sin(B/2)} = \frac{b}{\sin(B/2)}$$

$$\gamma = I_AI_B = \frac{x+y}{\sin(C/2)} = \frac{c}{\sin(C/2)}$$

so that

$$\alpha\beta\gamma = \frac{abc}{\sin(A/2)\sin(B/2)\sin(C/2)}$$

$$= abc\sqrt{\frac{bc}{(s-b)(s-c)}}\sqrt{\frac{ca}{(s-c)(s-a)}}\sqrt{\frac{ab}{(s-a)(s-b)}}$$

$$= abc\sqrt{\frac{(abc)^2}{((s-a)(s-b)(s-c))^2}} = \frac{(abc)^2}{(s-a)(s-b)(s-c)} = \frac{s(abc)^2}{\Delta^2}.$$

Now $\sqrt{\sigma(\sigma-\alpha)(\sigma-\beta)(\sigma-\gamma)}$ is the area of triangle $\triangle I_A I_B I_C$. This is also equal to

$$\frac{I_A I_B \times I_C C}{2} = \frac{I_A I_B \, I_C I_A \, \cos(A/2)}{2}$$

$$= \frac{(x+y)(x+z)\cos(A/2)}{2\sin(C/2)\sin(B/2)}$$

$$= \frac{cb}{2}\sqrt{\frac{s(s-a)}{bc}}\sqrt{\frac{ab}{(s-a)(s-b)}}\sqrt{\frac{ca}{(s-c)(s-a)}}$$

$$= \frac{bc}{2}\sqrt{\frac{sa^2}{(s-a)(s-b)(s-c)}} = \frac{abcs}{2\Delta}.$$

Hence,

$$\frac{\alpha\beta\gamma}{\sqrt{\sigma(\sigma-\alpha)(\sigma-\beta)(\sigma-\gamma)}} = \frac{s(abc)^2}{\Delta^2}\cdot\frac{2\Delta}{sabc} = \frac{2abc}{\Delta} = \frac{2abc}{1}\cdot\frac{4R}{abc} = 8R.$$

Problem 11.49. *A parabola has focus F and directrix l.*
The chord PQ passes through the focus F. L and M lie on l and satisfy $LP \perp l$ and $MQ \perp l$. Suppose that PN intersects LM at R.

(1) Prove that $\angle LFM$ is a right angle;
(2) Prove that R is the mid-point of LM.

Solution. We need a diagram (not here drawn to true scale): draw LN parallel to PF with N on the axis of the parabola
Since $PLNF$ is a parallelogram, and since $PF = PL$, we have that $\triangle PFL$ and $\triangle FLN$ are congruent isosceles triangles. This means that LF bisects $\angle PFN$. Similarly MF bisects $\angle QFN$. Hence, $\angle LFM$ is a right angle.
The tangent to the parabola at P must bisect $\angle LPF$. Thus, PN is the tangent to the parabola at P. Since $PLNF$ is a rhombus, PN is perpendicular to LF. Also, LF must intersect the y–axis at the mid-point of LF and PN, which we call T.

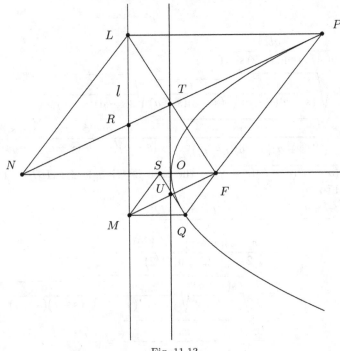

Fig. 11.13

Similarly, FM and SQ bisect one another on the y–axis at a point called U. We now easily see that TU is half the distance between PL and QM, and so $TU = \frac{1}{2}LM$.

Since PL is parallel to FM, we now see that $LR = RM = \frac{1}{2}LM$.

By a similar argument, we see that QS passes through R, and we obtain that $FTRU$ is a rectangle.

Problem 11.50. *The parabola* $y^2 = 4px$ *has focus* F *and directrix* $\mathcal{L}$.

(1) Write down the co-ordinates of F.

(2) Write down the equation of $\mathcal{L}$.

    Let P *be any point on the parabola (other than the nose) in the first quadrant.*

    The focal chord PF *meets the parabola again at* Q.

(3) Determine the co-ordinates of Q.

    U *is the point of* $\mathcal{L}$ *such that* $UP = PF$ *and* V *is the point on* $\mathcal{L}$ *such that* $VQ = QF$.

(4) Show that UF *is perpendicular to* VF.

Solution. The standard equation of a parabola is given in the text, so that

(1) $F = (0, p)$,

(2) $\mathcal{L}$ is $x = -p$.

(3) Suppose that $P = (\beta, \sqrt{4p\beta})$. The equation of PF is

$$\frac{y-p}{x-0} = \frac{\sqrt{4p\beta}-p}{\beta} \qquad \text{or} \qquad y = p + x\frac{\sqrt{4p\beta}-p}{\beta}.$$

We solve this with the equation of the parabola to get:

$$\left(p + x\frac{\sqrt{4p\beta}-p}{\beta}\right)^2 = 4px,$$

which reduces to

$$(x-\beta)\left(p\beta x - 4\beta - p + 4\sqrt{p\beta}\right) = 0.$$

Thus, the coordinates of Q are given by

$$\left(\frac{4\beta + p - \sqrt{p\beta}}{p\beta}, \frac{-2\sqrt{4\beta + p - 4\sqrt{\beta p}}}{\sqrt{\beta}}\right).$$

We could simplify matters if we were to take the coordinates of P in parametric form as $(pt^2, 2pt)$.

The equation of the focal chord is then

$$\frac{y-p}{x} = \frac{2pt - p}{pt^2} = \frac{2t-1}{t^2}.$$

We solve this with the equation of the parabola to get:

Problem 11.51. (This is Problem 5.2 on page 88.) *Suppose that P is any point of an ellipse with foci F and F'.*
Show that $PF + PF'$ is a constant.
For an ellipse in standard position, calculate the value of $PF + PF'$. See Fig 11.14 on the next page.

Solution. We use the standard coordinate system. So F is $(ae, 0)$, F' is $(-ae, 0)$ and P is (x, y). The directrices are $l : x = a/e$ and $l' : x = -a/e$. Draw lines from P perpendicular to the directrices, meeting them at N and N'. Join FPF'.
From the definition of an ellipse we have that $PF = e\,PN$.
Similarly, $PF' = e\,PN'$.
Hence, $PF + PF' = e(PN + PN') = e \times$ (the distance between the directrices) $= e \times 2\frac{a}{e} = 2a$, which is a constant.
It is, in fact, equal to the length of the major axis of the ellipse.

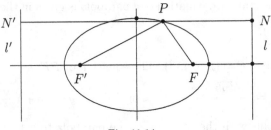

Fig. 11.14

Problem 11.52. (This is Problem 5.3 on page 88.) *Suppose that PQ and LM are two parallel chords of an ellipse.*
Suppose that R and N are the mid-points of PQ and LM, respectively.
Show that the line RN passes through the centre of the ellipse.

Solution. If the line is parallel to the y–axis, the result is easy to obtain. Thus, we may take the line to be $y = mx + c$.

The standard ellipse has equation $\frac{x^2}{a^2} + \frac{y^2}{b^2} = 1$. Substituting into the equation of the ellipse gives $\frac{x^2}{a^2} + \frac{(mx+c)^2}{b^2} = 1$. We solve this to get the x–coordinates of the points of intersection, and then substitute into the equation of the line to get the y–coordinates of the points of intersection. These are

$$x = \frac{a(b\sqrt{a^2m^2 + b^2 - c^2} - acm)}{a^2m^2 + b^2}$$

$$y = c + \frac{am(b\sqrt{a^2m^2 + b^2 - c^2} - acm)}{a^2m^2 + b^2}$$

and

$$x = -\frac{a(b\sqrt{a^2m^2 + b^2 - c^2} + acm)}{a^2m^2 + b^2}$$

$$y = c - \frac{am(b\sqrt{a^2m^2 + b^2 - c^2} + acm)}{a^2m^2 + b^2}.$$

To find the mid-point, we take the average of these, and to show that all such points lie on a line through the origin, we consider the ratio of the y–coordinate to the x–coordinate; that is, the slope of the line from the mid-point to the origin. This is

$$\frac{\frac{-am(b\sqrt{a^2m^2+b^2-c^2}+acm)}{(a^2m^2+b^2)} + \frac{(am(b\sqrt{a^2m^2+b^2-c^2}-acm)}{(a^2m^2+b^2)+c)}}{\frac{-a(b\sqrt{a^2m^2+b^2-c^2}+acm)}{a^2m^2+b^2} + \frac{a(b\sqrt{a^2m^2+b^2-c^2}-acm)}{a^2m^2+b^2}}$$

which simplifies to $-\frac{b^2}{a^2m}$.

Since this is independent of c, the intercept of the line with the y–axis, all mid-points lie on the same line.

Problem 11.53. (This is Problem 5.4 on page 90.) *Suppose that F and F' are the foci of a hyperbola and that P is any point on the hyperbola.*
Calculate the set of values taken by $|PF| - |PF'|$. ($|PF|$ means the length of the line segment PF.)

Solution. We start with a diagram:

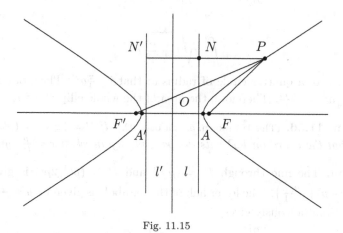

Fig. 11.15

From the definition of a hyperbola, $PF = ePN$ and $PF' = ePN'$.

Thus, $PF - PF' = e(PN - PN') = -e\,NN' = -e\,2\frac{a}{e} = -a$, which is constant.

Compare with Problem 10.1.

Problem 11.54. (This is Problem 5.5 on page 90.) *Calculate the eccentricity for the rectangular hyperbola $xy = 4$.*

Solution. The actual equation is irrelevant — so we make take any rectangular hyperbola. We take $\frac{x^2}{a^2} - \frac{y^2}{a^2} = 1$.

The asymptotes are give by $\frac{x^2}{a^2} - \frac{y^2}{a^2} = 0$, or $y = \pm x$.

Thus, $a^2 = b^2 = a^2(e^2 - 1)$, so that $e = \sqrt{2}$.

Problem 11.55. (This is Problem 5.7 on page 92.) *Find the area of an ellipse.*
(*Take an ellipse in standard form.*)

Solution. The standard ellipse has equation $\frac{x^2}{a^2} + \frac{y^2}{b^2} = 1$.
If we find the area in the first quadrant, symmetry allows us to multiply
that area by 4 to get the required area.
In the upper half plane, the equation of the standard ellipse is $y = b\sqrt{1 - \frac{x^2}{a^2}}$. Therefore, standard Calculus techniques tell us that the area in
the first quadrant bounded by the axes and the ellipse is given by

$$\int_0^a y\,dx = \int_0^a b\sqrt{1 - \frac{x^2}{a^2}}\,dx \ .$$

Now,

$$\int_0^a a\sqrt{1 - \frac{x^2}{a^2}}\,dx$$

is the area of a quarter circle of radius a; that is, $\frac{\pi}{4}a^2$. Thus, our area is
then $\frac{b}{a}\frac{\pi}{4}a^2 = \frac{\pi ab}{4}$. Therefore, the area of the whole ellipse is πab.

Problem 11.56. (This is Problem 5.6 on page 92.) *If $P = (x,y) = (pt^2, 2pt)$,
prove that the chord on FP cuts the parabola again when $x = \frac{p}{t^2}$, $y = \frac{2p}{t}$.*

Solution. The line through $F = (p,0)$ and $P = (pt^2, 2pt)$ is given by
$y = (x-p)\left(\frac{2t}{t^2-1}\right)$. The lower half of the parabola is given by $y = -2\sqrt{px}$.
Solving simultaneously gives

$$x\left(\frac{2t}{t^2-1}\right) - 2\sqrt{px} - p\left(\frac{2t}{t^2-1}\right) = 0 \ .$$

This factors to

$$\left(\frac{2t}{t^2-1}\right)\left(\sqrt{x} - \frac{\sqrt{p}}{t}\right)\left(\sqrt{x} + \sqrt{pt}\right) \ ,$$

yielding the solution.

Problem 11.57. (This is Problem 5.8 on page 93.) *Find the conditions of the
numbers a, b, c, f, g and h for the general equation of second degree*

$$ax^2 + 2hxy + by^2 + 2gx + 2fy + c = 0$$

to be a circle.

Solution.

$$h = 0 \ , \qquad a = b \neq 0 \qquad f^2 + g^2 > ac \ .$$

Problem 11.58. (This is Problem 5.9 on page 94) *For what values of h does the general equation of second degree*

$$ax^2 + 2hxy + by^2 + 2gx + 2fy + c = 0$$

represent a pair of lines?

Solution.

$$h = \frac{f * g - \sqrt{ac(bc - f^2) - bcg^2 + f^2 g^2}}{c},$$

$$h = \frac{\sqrt{ac(bc - f^2) - bcg^2 + f^2 g^2} + fg}{c}.$$

Problem 11.59. (This is Problem 5.10 on page 94.) *Suppose that A and B are two distinct fixed points.*
Find the locus of the point P which has the property that

$$\angle PAB - \angle PBA$$

is constant.
What relation does the line segment AB have with this locus?

Solution. Without loss of generality, let $A = (1, 0)$ and $B = (-1, 0)$.

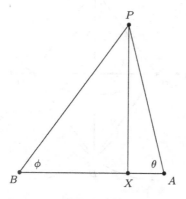

Fig. 11.16

Drop a perpendicular from P to AB, to meet at X. Let $\angle PAB = \theta$ and $\angle PBA = \phi$. Then $\tan(\theta) = \frac{PX}{XA} = \frac{y}{1-x}$ and $\tan(\phi) = \frac{PX}{BX} = \frac{y}{1+x}$.

Thus,

$$\tan(\theta - \phi) = \frac{\tan(\theta) + \tan(\phi)}{1 - \tan(\theta)\tan(\phi)} = \frac{\frac{y}{1-x} + \frac{y}{1+x}}{1 - \frac{y}{1-x}\frac{y}{1+x}}$$

$$= \frac{\frac{y}{1-x^2}(1 - x - 1 - x)}{1 + \frac{y^2}{1-x^2}}$$

$$= \frac{-2xy}{1 - x^2 - y^2} \ .$$

This is a constant, so that we have $1 - x^2 - y^2 = -2cxy$. If $c \neq \pm 1$, this is a rectangular hyperbola, with AB as a chord. If $c = \pm 1$, then we get a pair of parallel lines, through the points A and B.

Problem 11.60. (This is Problem 6.1 on page 99.) *"We now consider a line AB and a point C not on the line.*
We begin by drawing an arc of a circle, centre C to intersect AB at two points P and Q. With centres P and Q, draw arcs with the same radius to intersect on the other side of AB from C. Suppose that these arcs intersect at D. Then CD is perpendicular to AB."
In the above construction, prove that $CD \perp AB$.

Solution. Join C and D to P and Q.

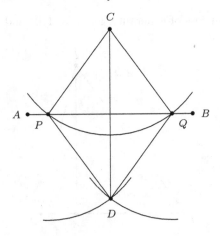

Fig. 11.17

Note that $CPDQ$ is a rhombus. We know that the diagonals of a rhombus are perpendicular to one another.

Problem 11.61. (This is Problem 6.3 on page 102.)
Show that $\cos(36°) = \frac{1+\sqrt{5}}{4}$.

Solution. First, using DeMoivre's Theorem and Pythagoras' Theorem, we obtain that

$$\begin{aligned}
\cos(5x) &= \cos^5(x) - 10\cos^3(x)\sin^2(x) + 5\cos(x)\sin^4(x) \\
&= \cos^5(x) - 10\cos^3(x)\left(1 - \cos^2(x)\right) + 5\cos(x)\left(1 - \cos^2(x)\right)^2 \\
&= \cos^5(x) - 10\cos^3(x)\left(1 - \cos^2(x)\right) \\
&\quad + 5\cos(x)\left(1 - 2\cos^2(x) + \cos^4(x)\right) \\
&= \cos(x)\left(16\cos^4(x) - 20\cos^2(x) + 5\right) \ .
\end{aligned}$$

Thus, setting $x = \pi/5 = 36°$, we have

$$-1 = \cos(\pi) = \cos(\pi/5)\left(16\cos^4(\pi/5) - 20\cos^2(\pi/5) + 5\right) \ .$$

It is now easy to verify that $\frac{1+\sqrt{5}}{4}$ satisfies this equation.
However, this does not **prove** the result — it merely shows that it may be a solution!
From $\cos(36°) = \frac{1+\sqrt{5}}{4}$, we obtain that

$$\cos(72°) = 2\cos^2(36°) - 1 = \frac{2}{16}(6 + 2\sqrt{5}) - 1 = \frac{\sqrt{5} - 1}{4} \ .$$

Double again!

$$\cos(144°) = 2\cos^2(72°) - 1 = \frac{2}{16}(6 - 2\sqrt{5}) - 1 = -\frac{\sqrt{5} + 1}{4} \ ,$$

and this does prove the result.

Problem 11.62. (This is Problem 6.4 on page 103.) *Show that the following is a method to construct a regular pentagon:*

Draw a circle, centre O, into which the circle is to be inscribed.
Draw any diameter AOB.
Bisect OB to get C.
Find D on the circle so that OD is perpendicular to AOB.
With centre C and radius CD, draw an arc to intersect AB internally at E.
With centre D and radius DE, draw an arc to intersect the arc of the circle, AD, at F.
Then DF is one side of the regular pentagon required.

Solution. We start by noting that the angle at the centre of a regular pentagon is $\frac{2\pi}{5} = 70°$. From the previous example, we have that $\cos(72°) = \frac{\sqrt{5}-1}{4}$. Suppose, for convenience, that the radius of the circle is 2. Then, from triangle OPQ, the side of the regular pentagon is $\sqrt{10 - 2\sqrt{5}}$.

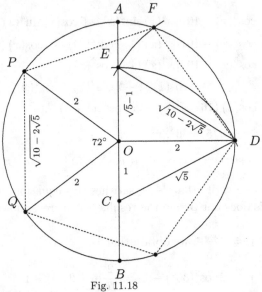

Fig. 11.18

Since $OC = 1$ and $OD = 2$, we get that $DC = \sqrt{5}$. Thus, $OE = \sqrt{5} - 1$, and further, $FD = DE = \sqrt{10 - 2\sqrt{5}}$.

Problem 11.63. (This is Problem 6.5 on page 104.) *Show that the following is a method to construct a regular pentagon:*

We need to find a point G such that

$$OG = \frac{\sqrt{5} - 1}{4}$$

times the radius of the given circle.

Draw a circle, centre O, into which the circle is to be inscribed.
Let OA be any radius and extend it to B so that $OA = AB$.
Draw BC perpendicular to OB with $BC = OA$.
Note that $OC = \sqrt{5} \times OA$.
Let OC intersect the circle at D.

Find E on the line segment OC such that CE = OA.
Bisect OE to find F. Bisect OF to find G.
Draw HGJ perpendicular to OC, where H and J lie on the given circle.
Then H, D and J are three vertices of the regular pentagon.

Solution. As in the last problem, we take a circle of radius 2 and need to show that $HD = DJ$ is the side of the regular pentagon; that is $HD = \sqrt{10 - 2\sqrt{5}}$.

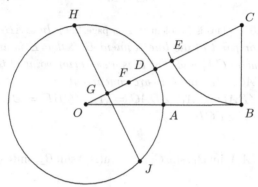

Fig. 11.19

We know that $OC = 2\sqrt{5}$ and $CE = 2$, giving $OE = 2(\sqrt{5} - 1)$. Thus, $OG = \frac{\sqrt{5}-1}{2}$, giving $DG = 2 - \left(\frac{\sqrt{5}-1}{2}\right) = \frac{5-\sqrt{5}}{2}$.
Using right triangle OGH, we have $HG^2 = OH^2 - OG^2 = 4 - OG^2$. Using right triangle DGH, we have

$$HD^2 = HG^2 + GD^2 = 4 - OG^2 + GD^2 = 10 - 2\sqrt{5} .$$

That $HD = DJ$ follows by symmetry.

Problem 11.64. (This is Problem 10.1 on page 129.) *In $\triangle ABC$, we have $\angle BAC = 90°$. Let D lie on BC so that $AD \perp BC$. Suppose that $AD = 1$. Show that $BC = AB \times AC$. See Fig. 11.20.*

Solution. First, we note that $\angle BAC = \angle BDA = \angle ADC$ and that $\angle BAD = 90° - \angle DAC = \angle DCA = \angle BCA$.
Similarly, $\angle ABC = \angle ABD = 90° - \angle BAC = \angle DAC$.
Hence, $\triangle ABC$, $\triangle CAD$ and $\triangle DBA$ are all similar.
Therefore, $BC : AC : AB = AC : DC : AD = AB : AD : BD$.
Thus, $\frac{BC}{AB} = \frac{AC}{AD}$, yielding $AD \times BC = AB \times AC$. Since we are given that $AD = 1$, the result follows.

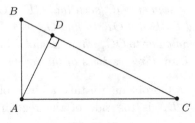

Fig. 11.20

Problem 11.65. (This is Problem 10.2 on page 129.) *In $\triangle ABC$, suppose that D is an interior point of the line segment BC, that E is an interior point of the line segment CA, and that F is an interior point of the line segment AB such that AC, BE and CF are concurrent.*
Suppose that $\angle BAD = A_1$, $\angle DAC = A_2$, $\angle CBE = B_1$, $\angle EBA = B_2$, $\angle ACF = C_1$ and $\angle FCB = C_2$.
Prove that

$$\sin(A_1)\sin(B_1)\sin(C_1) = \sin(A_2)\sin(B_2)\sin(C_2) \quad .$$

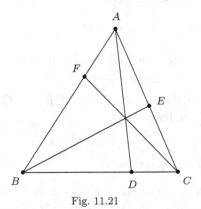

Fig. 11.21

Solution. From the Sine Rule applied to $\triangle ABD$, we have that $\frac{\sin(A_1)}{BD} = \frac{\sin(B)}{AD}$.
Using this and similar results from other triangles, we obtain

$$\sin(A_1) = \sin(B) \times \frac{BD}{AD} \qquad \sin(A_2) = \sin(C) \times \frac{DC}{AD}.$$

$$\sin(B_1) = \sin(C) \times \frac{CE}{CF} \qquad \sin(B_2) = \sin(A) \times \frac{AE}{CF}.$$

$$\sin(C_1) = \sin(A) \times \frac{AF}{BE} \qquad \sin(C_2) = \sin(B) \times \frac{BF}{EC}.$$

Therefore,

$$\sin(A_1)\sin(B_1)\sin(C_1) = \frac{\sin(A)\sin(B)\sin(C)}{AD\cdot BE\cdot CF}\times(AF\cdot BD\cdot CE) \ ,$$

$$\sin(A_2)\sin(B_2)\sin(C_2) = \frac{\sin(A)\sin(B)\sin(C)}{AD\cdot BE\cdot CF}\times(FB\cdot DC\cdot EA) \ .$$

From Ceva's Theorem, we know that $\frac{AF}{FD}\cdot\frac{BD}{DC}\cdot\frac{CE}{EA}=1$. The result now follows.

Problem 11.66. (This is Problem 10.3 on page 130.) *Given convex quadrilateral ABCD, extend AB, BC, CD and DA to W, X, Y and Z, respectively, such that $AW = 2AB$, $BX = 2BC$, $CY = 2CD$ and $DZ = 2DA$. Show that $[WXYZ] = 5\times[ABCD]$.*

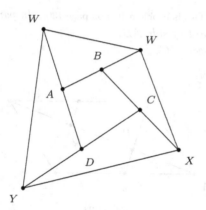

Fig. 11.22

Solution. We have

$$[AVW] = \frac{1}{2}AW\cdot AZ\sin(W\widehat{A}Z) = AB\cdot DA\sin(D\widehat{A}B) \ .$$

Similarly,

$$[BWX] = BC\cdot AB\sin(A\widehat{B}C) \ ,$$

$$[CXY] = CD\cdot BC\sin(B\widehat{C}D)$$

and

$$[DYC] = DA\cdot CD\sin(C\widehat{D}A) \ .$$

From joining AC, we have

$$[ABCD] = \frac{1}{2}AB \cdot CD\sin(A\widehat{B}C) + \frac{1}{2}CD \cdot DA\sin(C\widehat{D}A) \ .$$

From joining BD, we have

$$[ABCD] = \frac{1}{2}BC \cdot CD\sin(B\widehat{C}D) + \frac{1}{2}DA \cdot AB\sin(D\widehat{A}B) \ .$$

Thus, $4[ABCD] = [AZW] + [BWX] + [CXY] + [DYZ]$, and further, $4[ABCD] = [WXYZ]$.

Problem 11.67. (This is Problem 10.4 on page 130.) *In parallelogram $ABCD$, let E be the mid-point of side AB.*
Prove that DE trisects AC.

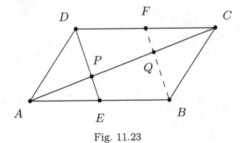

Fig. 11.23

Solution. Let F be the mid-point of DC. Join BF.
Since $AE = \frac{1}{2}AB = \frac{1}{2}CD = CF$, $AD = CB$ (both using opposite sides of a parallelogram being equal), and $\angle DAE = \angle BCF$ (opposite angles of a parallelogram being equal), we have $\triangle ADE \equiv \triangle CBF$.
Hence, $\angle ADE = \angle CFB$, and thus, $DE \parallel FB$.
In $\triangle ABQ$, $\triangle APE$, we have $BQ \parallel EP$ and $AE = \frac{1}{2}AB$.
Hence, $AP = \frac{1}{2}AQ$.
By symmetry, $AP = CQ$, and hence, $AP = \frac{1}{3}AC$.

Problem 11.68. (This is Problem 10.5 on page 130.) *In triangle ABC, let E and D lie on BC such that CE = ED = DB. Let F be the mid-point of AC and G be the mid-point of AB. Suppose that EG and FD meet at H. Find the value of* $\frac{EH}{HG}$.

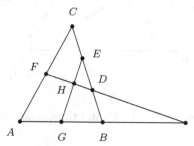

Fig. 11.24

Solution. Consider $\triangle ABC$ with transversal KDF: we have
$$\frac{AK}{KB} \cdot \frac{BD}{DC} \cdot \frac{CF}{FA} = -1 .$$
Hence, $\frac{AK}{KB} \cdot \frac{1}{2} \cdot (-1) = -1$, so that $AK = 2AB = 4GB$.
Consider $\triangle GBE$ with transversal KDH: we have
$$\frac{GK}{KB} \cdot \frac{BD}{DE} \cdot \frac{EH}{HG} = -1 .$$
Hence, $-\frac{3}{2} \cdot (1) \cdot \frac{EH}{HG} = -1$, so that $\frac{EH}{HG} = \frac{2}{3}$.
Alternatively: if we set up a coordinate system with the origin at A, with K as $(16, 0)$ and C as $(5, 9)$, we obtain that:
the equation of FD is $y = -\frac{x}{3} + \frac{16}{3}$;
the equation of GE is $y = 3x - 12$.
Solving these simultaneously gives $H = (5.2, 3.6)$, and the ratio is easily calculated.

Problem 11.69. (This is Problem 10.6 on page 130.) *On triangle ABC, construct similar isosceles triangles ABC' and CAB' outwards on bases AB and CA, respectively, and isosceles triangle BCA' inwards on base BC.*
Prove that $AB'A'C'$ is a parallelogram.

Solution. Note that $\angle A'BC = \angle C'BA$, so that $\angle C'BA' = \angle ABC$. Since $C'B : BA' = AB : BC$, it follows that $\triangle C'BA'$ is similar to $\triangle ABC$. Hence, $C'A' = AB'$.

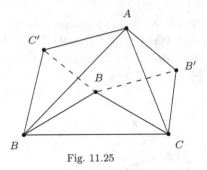

Fig. 11.25

By a similar argument, $AC' = B'A'$, and thus, quadrilateral $AB'A'C'$ is a parallelogram.

Problem 11.70. (This is Problem 10.7 on page 131.) *The sides AB, BC, CD and DA of quadrilateral $ABCD$ are divided by the points E, F, G and $H such that*

$$\frac{AE}{EB} = \frac{CF}{FB} = \frac{CG}{GD} = \frac{DH}{HA} \ .$$

Show that $EFGH$ is a parallelogram.

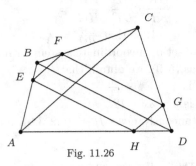

Fig. 11.26

Solution. Join BD.

In $\triangle ABD$, we have $AE : EB = AH : HD$, therefore, $EH \parallel BD$.

Similarly, from $\triangle CBD$, we get $FG \parallel BD$. Hence, $EH \parallel FG$.

Join AC. A similar argument shows that $EF \parallel HG$.

Hence, $EFGH$ is a parallelogram.

Problem 11.71. (This is Problem 10.8 on page 131.) *In triangle ABC, points D and E lie on sides BC and CA, respectively, so that $\frac{BD}{DC} = 3$ and $\frac{CE}{EA} = \frac{2}{3}$. Suppose that AD and BE meet at P.*
Find the value of $\frac{BP}{PE}$.
Find the value of $\frac{AP}{PD}$.

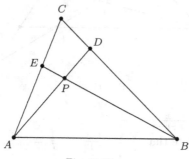

Fig. 11.27

Solution. Consider $\triangle BEC$ with transversal APD:
we have $\frac{BP}{PE} \cdot \frac{EA}{AC} \cdot \frac{CD}{DB} = -1$.
Thus, $\frac{BP}{PE} \cdot \frac{-3}{5} \cdot \frac{1}{3} = -1$, so that $\frac{BP}{PE} = 5$.
Consider $\triangle ACD$ with transversal BPE: we have $\frac{AP}{PD} \cdot \frac{DB}{BC} \cdot \frac{CE}{EA} = -1$.
Thus, $\frac{AP}{PD} \cdot \frac{-3}{4} \cdot \frac{2}{3} = -1$, so that $\frac{AP}{PD} = 2$.

Problem 11.72. (This is Problem 10.9 on page 131.) *In triangle ABC, points E and F lie on sides CA and AB, respectively, so that $\frac{AE}{EC} = 4$ and $\frac{AF}{FB} = 1$. Suppose that D lies on BC and AD meets EF at G. Suppose that $\frac{AG}{GD} = \frac{3}{2}$. See Fig. 11.28.*
Find the value of $\frac{BD}{DC}$.

Solution. The ratios given are indicated by the numbers in circles. The unknown ratios of $BD : DC : CH$ are given by $x : y : z$.
Consider $\triangle ABD$ with transversal FGH: we have $\frac{AF}{FB} \cdot \frac{BH}{HD} \cdot \frac{DG}{GA} = -1$.
Thus, $1 \cdot \frac{BH}{HD} \cdot \frac{2}{3} = -1$, giving $\frac{BH}{HD} = \frac{3}{2}$.
Consider $\triangle ABC$ with transversal FEH: we have $\frac{AE}{EC} \cdot \frac{CH}{HB} \cdot \frac{BF}{FA} = -1$.
Thus, $4 \cdot \frac{CH}{HB} \cdot 1 = -1$, giving $\frac{CH}{HB} = \frac{1}{4}$.
Therefore,
$$\frac{y+z}{x+y+z} = \frac{2}{3}, \qquad \frac{z}{x+y+z} = \frac{1}{4}.$$
Thus, $\frac{y}{x+y+z} = \frac{2}{3} - \frac{1}{4} = \frac{7}{12}$.

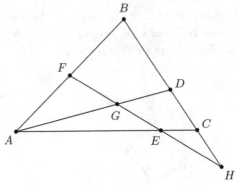

Fig. 11.28

We now deduce that $\frac{x}{x+y+z} = \frac{1}{3}$.

Hence, we can calculate that $\frac{x}{y} = \frac{1/3}{7/12} = \frac{4}{7} = \frac{BD}{DC}$.

Problem 11.73. (This is Problem 10.10 on page 131.) *On the sides of an arbitrary parallelogram $ABCD$, squares are constructed lying exterior to the parallelogram. Denote the centres of these squares by P_1, P_2, P_3 and P_4.*
Prove that quadrilateral $P_1P_2P_3P_4$ is a square.

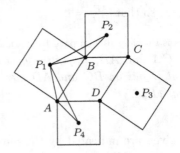

Fig. 11.29

Solution. If $ABCD$ is a square, the proof is very easy. Assume that $ABCD$ is not a square. Assume, without loss of generality that $\angle ABC < 90°$. Join AP_4, AP_1, BP_1 and BP_2.
By symmetry, $P_1P_2 = P_4P_3$ and $P_1P_4 = P_2P_3$. Hence, $P_1P_2P_3P_4$ is a parallelogram.
We have $\angle P_2BC = \angle P_1BA = \angle P_1AB = \angle P_4AD = 45°$.

Also, $\angle P_2BP_1 + \angle P_2BC + \angle P_1BA + \angle ABC = 360°$.
Therefore,

$$\angle P_2BP_1 = 360° - 45° - 45° - \angle ABC$$
$$= 270° - \angle ABC$$
$$= 270° - (180° - \angle BAD) = 90° + \angle BAD \ ;$$

$$\angle P_1AP_4 = \angle P_1AB + \angle BAD + \angle DAP_4$$
$$= 45° + \angle BAD + 45° = 90° + \angle BAD$$
$$= \angle P_2BP_1 \ .$$

Since $P_2B = P_4A$ and $P_1B = P_1A$, we deduce that $\triangle P_1BP_2 \equiv \triangle P_1AP_4$.
Hence, $P_1P_2 = P_1P_4$, and further, $P_1P_2P_3P_4$ is a rhombus.
Also, we have that $\angle P_2P_1B = \angle P_4P_1A$. Since $\angle BP_1A = 90°$, we have
$\angle P_2P_1P_4 = 90°$. Hence, $P_1P_2P_3P_4$ is a square.

Problem 11.74. (This is Problem 10.11 on page 131.) *On the sides of an
arbitrary convex quadrilateral ABCD, equilateral triangles ABM_1, BCM_2,
CDM_3 and DAM_4 are constructed so that ABM_1 and CDM_3 are exterior
to the quadrilateral and BCM_2 and DAM_4 are drawn "inwards" to the
quadrilateral.*
Show that the quadrilateral $M_1M_2M_3M_4$ is a parallelogram.

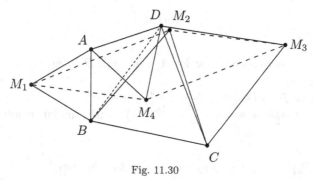

Fig. 11.30

Solution. For $\triangle M_1AM_4$ and $\triangle BAD$, we have $M_1A = BA$, $AM_4 = AD$
and $\angle M_1AM_4 = \angle BAD$ (since $\angle M_1AB = \angle M_4AD = 60°$). Hence,
$M_4M_1 = DB$.
By a similar argument using $\triangle M_3CM_2$ and $\triangle BCD$, we get $M_2M_3 = DB$.
Hence, $M_4M_1 = M_2M_3$. A similar argument shows that $M_4M_3 = M_1M_2$.
Thus, $M_1M_2M_3M_4$ is a parallelogram.

Problem 11.75. (This is Problem 10.12 on page 131.) *On the sides of an arbitrary convex quadrilateral ABCD, squares are constructed all lying external to the quadrilateral, with centres P_1, P_2, P_3 and P_4. Show that $P_1P_3 = P_2P_4$ and that P_1P_3 is perpendicular to P_2P_4.*

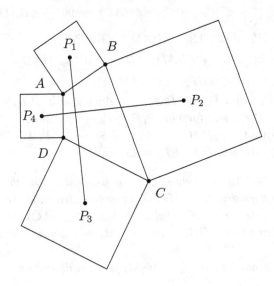

Fig. 11.31

Solution. We use coordinates: let $A = (a_1, a_2)$, $B = (b_1, b_2)$, $C = (a_3, a_4)$ and $D = (b_3, b_4)$.

Then we have $P_1 = \left(\frac{a_1+b_1}{2} + \frac{b_2-a_2}{2}, \frac{a_2+b_2}{2} - \frac{b_1-a_1}{2} \right)$
and $P_3 = \left(\frac{a_3+b_3}{2} + \frac{b_4-a_4}{2}, \frac{a_4+b_4}{2} - \frac{b_3-a_3}{2} \right)$, with similar results for P_2 and P_4.

Thus,

$$4(P_1P_3)^2 = (a_1 + b_1 + b_2 - a_2 - a_3 - b_3 - b_4 + a_4)^2$$
$$+ (a_2 + b_2 - b_1 + a_1 - a_4 - b_4 + b_3 - a_3)^2$$
$$= (-a_3 - b_3 + b_2 - a_2 + a_1 + b_1 - b_4 + a_4)^2$$
$$+ (a_2 + b_2 + b_3 - a_3 - a_4 - b_4 - b_1 + a_1)^2$$
$$= (a_2 + b_2 + b_3 - a_3 - a_4 - b_4 - b_1 + a_1)^2$$
$$+ (-a_3 - b_3 + b_2 - a_2 + a_1 + b_1 - b_4 + a_4)^2$$
$$= 4(P_2P_4)^2 .$$

Thus, we have $P_1P_3 = P_2P_4$.

Also

$$m(P_1P_3) = \frac{a_2 + b_2 - b_1 + a_1 - a_4 - b_4 + b_3 - a_3}{a_1 + b_1 + b_2 - a_2 - a_3 - b_3 - b_4 + a_4}$$

$$= \frac{-1}{\dfrac{a_3 + b_3 - b_2 + a_2 - a_1 - b_1 + b_4 - a_4}{a_2 + b_2 + b_3 - a_3 - a_4 - b_4 - b_1 + a_1}}$$

$$= \frac{-1}{m(P_2P_4)} \; ,$$

showing that P_1P_3 and P_2P_4 are perpendicular.

Problem 11.76. (This is Problem 10.13 on page 131.) *Let ABC be an acute angled triangle. Construct squares externally on the sides, and extend the altitudes to meet the opposite sides of the squares, giving rectangles with areas as indicated on the diagram:*

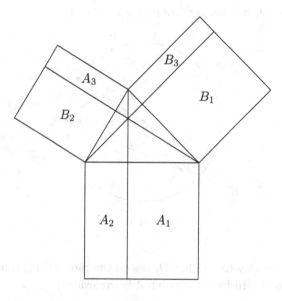

Fig. 11.32

Show that, for $i = 1$, 2, 3, $A_i = B_i$.

Solution. We have $A_1 = QR \cdot DR = QR \cdot HR\cos(H\widehat{R}D)$ and $B_1 = PR \cdot RE = PR \cdot HR\cos(H\widehat{R}E)$. Therefore,

$$
\begin{aligned}
\frac{A_1}{B_1} &= \frac{QR\cos(H\widehat{R}D)}{PR\cos(H\widehat{R}E)} \\
&= \frac{QR\cos(F\widehat{R}Q)}{PR\cos(F\widehat{R}P)} \\
&= \frac{FR}{FR} = 1 \ .
\end{aligned}
$$

The others follow similarly.

Problem 11.77. (This is Problem 10.14 on page 132.) *Let A be a given point and let P_i, $i = 1, 2, \ldots, n$, $n \geq 7$, be the vertices of a regular n–gon. Let the points Q_i be given by*

$$
\overrightarrow{AQ_i} = \overrightarrow{AP_i} + \overrightarrow{P_{i+1}P_{i+2}} \ , \qquad i = 1, 2, \ldots, n
$$

$(P_{n+i} = P_i, i = 1, 2)$.
Prove that the Q_i are the vertices of a regular n–gon.

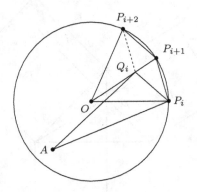

Fig. 11.33

Solution. It is easy to see that Q_i lies on the radius OP_{i+1}, and is a fixed distance from O. (In fact, the point A is irrelevant.)

Problem 11.78. (This is Problem 10.15 on page 132.) *Given a circle with centre O and a chord AB, let C be any other point on AB. Join OC and extend it (in the direction $O \to C$) to meet the circle at D.*
Calculate the radius of the circle in terms of AC, BC and CD.

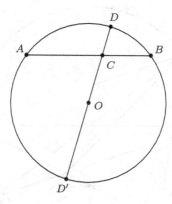

Fig. 11.34

Solution. Extend DCO to meet the circle again at D'. By the Intersecting Chords Theorem, we have $AC \cdot CB = DC \cdot CD' = CD(2r - CD)$. Hence, $2r = \frac{AC \cdot CB}{CD} + CD$, yielding $r = \frac{AC \cdot CB + CD^2}{2CD}$.

Problem 11.79. (This is Problem 10.16 on page 132.) *Given a circle with centre O and two fixed points on the circumference, A and B such that AB is not a diameter.*
Suppose that PQ is any diameter. Suppose that AP and BQ intersect at X. Find the locus of X as P moves once around the circle.

Solution. We first draw another position: when PQ is parallel to AB. Re-letter this position of PQ as LM. Depending on whether LM is PQ or QP, we get the points Z and Y on the locus.
Note that the determination of Z and Y determines that MA and LB are altitudes of isosceles $\triangle LMZ$ (isosceles since $LZ = MZ$).
Now,

$$\angle AXB = \angle PXQ = 180° - \angle XPQ - \angle XQP$$
$$= 180° - \angle APQ - \angle BQP$$
$$= 180° - \angle ALQ - \angle BMP \quad \text{(angles in same segment)}$$
$$= 180° - (\angle ALQ + \angle QLM) - (\angle BMP - \angle PML)$$
$$\text{(from rectangle } PMQL \text{ and diagonal } AM)$$
$$= 180° - \angle ALM - \angle BML$$
$$= \angle LZM = \angle AZM \ .$$

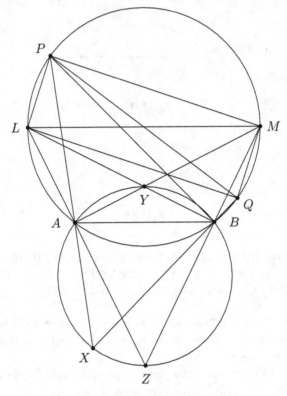

Fig. 11.35

But, $\angle AZM$ is fixed, being determined by the given circle and the given fixed chord; therefore, $\angle AXM$ is constant, yielding that X lies on the circle with ZY as diameter.

Problem 11.80. (This is Problem 10.17 on page 132.) *Given a circle and two fixed points on the circumference, A and B, let CD be a chord of fixed length and orientation. Suppose that AC and BD intersect at P.*
Find the locus of P as chord CD moves once around the circle.

Solution. First we deal with the case where the chord CD is such that it does not intersect AB and P is outside the given circle.
Note that $\angle ADB$ is of fixed size (angle subtended by chord AB, of fixed length). Note also that $\angle PAD$ is of fixed size (subtended by chord CD, of fixed length.
Hence, $\angle APB = \angle ADB - \angle PAD$ is of fixed size.

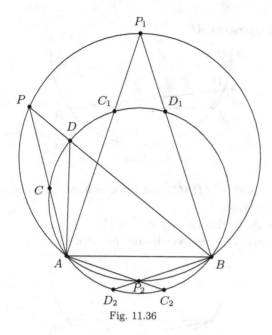

Fig. 11.36

Thus, P lies on a circle.

Let P_1 and P_2 be the positions of P when chord CD takes positions C_1D_1 and C_2D_2, which are parallel to chord AB. Then C_1C_2 and D_1D_2 are diameters of the given circle. This means that $\angle P_1AP_2$ and $\angle P_1BP_2$ are right angles.

Hence, P lies on the circle with diameter P_1P_2.

We now deal with the remaining cases to show that P lies on the same circle.

Case (a) : $C = A$. In this case, the line CD is the tangent to the circle at A.

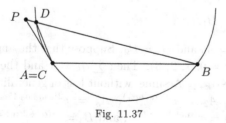

Fig. 11.37

Here, $\angle PAD = \angle DBC$ (by chord and tangent theorem) is fixed, and $\angle ADB$ is fixed. Hence, $\angle APB$ is fixed.

Case (b) : CD intersects AB.

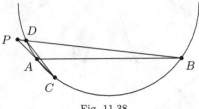

Fig. 11.38

Here, $\angle PAD = 180° - \angle DAC$, and thus, is fixed. Also, $\angle ADB$ is fixed. Hence, $\angle APB$ is fixed.

Case (c) : $D = A$. In this case, $A = D = P$.

Case (d) : AC and BD intersect inside the circle.

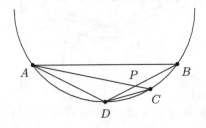

Fig. 11.39

The proof follows that same lines as the main proof.

Problem 11.81. (This is Problem 10.18 on page 133.) *Suppose that $A_0 A_1 A_2 A_3$ is a convex cyclic quadrilateral, and let $A_4 = A_0$. Suppose that B_k, $k = 0$, 1, 2 and 3, are points such that*

$$\text{arc } A_k B_k = \text{arc } B_k A_{k+1} \qquad k = 0, 1, 2, 3 \ .$$

Prove that $B_0 B_2 \perp B_1 B_3$.

Solution. Let $A_4 = A_0$ and $B_4 = B_0$. Suppose that the angle between A_k and A_{k+1} is $2\alpha_k$ ($k = 0, 1, 2, 3$). Then $\sum \alpha_k = \pi$ and the angle between B_k and B_{k+1} is $\alpha_k + \alpha_{k+1}$. Assume, without loss of generality, that $A_0 = 1$. Then $A_1 = e^{2i\alpha_0}$, $A_2 = e^{2i(\alpha_0+\alpha_1)}$, $A_3 = e^{2i(\alpha_0+\alpha_1+\alpha_2)}$, $B_0 = e^{i\alpha_0}$, $B_1 = e^{i(\alpha_0+2\alpha_1)}$, $B_2 = e^{i(\alpha_0+2\alpha_1+2\alpha_2)}$, and $B_3 = e^{i(\alpha_0+2\alpha_1+2\alpha_2+\alpha_3)}$.

We can now find the vectors for $B_2 B_0$ and $B_3 B_1$, and it is easy to verify that they are perpendicular to one another.

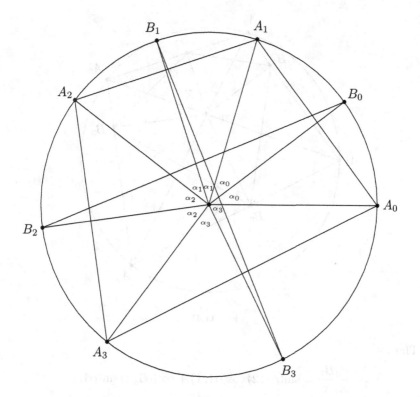

Fig. 11.40

Problem 11.82. (This is Problem 10.19 on page 133.) *Suppose that $A_0A_1A_2A_3$ is a convex cyclic quadrilateral, and let $A_4 = A_0$.*
Suppose that the diagonals A_0A_2 and A_1A_3 meet at X.
The feet of the perpendicular to A_kA_{k+1} is denoted by B_k, for $k = 0, 1, 2$ and 3. Let $B_4 = B_0$.
Prove that $|B_0B_1| + |B_2B_3| = |B_1B_2| + |B_3B_4|$.

Solution. Note all the cyclic quadrilaterals, since $XB_k \perp A_kA_{k+1}$ ($A_4 = A_0$). We use Ptolemy's Theorem:

$$B_0B_1 \cdot A_1X = B_1X \cdot A_1B_0 + B_0X \cdot A_1B_1 \ ,$$

yielding

$$B_0B_1 = \frac{B_1X \cdot A_1B_0 + B_0X \cdot A_1B_1}{A_1X}$$
$$= A_1B_0 \sin(\alpha_{1,2}) + A_1B_1 \sin(\alpha 1, 1) \ .$$

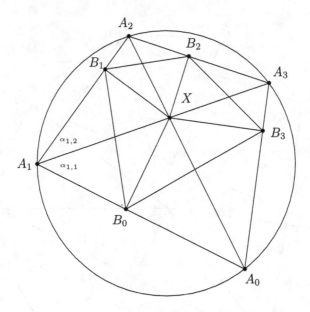

Fig. 11.41

Thus,

$$\frac{B_0 B_1}{A_1 X} = \sin(\alpha 1, 2)\cos(\alpha 1, 1) + \cos(\alpha 1, 2)\sin(\alpha 1, 1)$$

$$= \sin(\alpha 1, 2 + \alpha 1, 1) = \sin(A_0 \widehat{A_1} A_2) \ .$$

Thus, $B_0 B_1 = A_1 X \sin(A_0 \widehat{A_1} A_2)$. Similarly, $B_2 B_3 = A_3 X \sin(A_0 \widehat{A_3} A_2)$.
But $\sin(A_0 \widehat{A_1} A_2) = \sin(A_0 \widehat{A_3} A_2)$.
Hence, $B_0 B_1 + B_2 B_3 = A_1 A_3 \sin(A_0 \widehat{A_1} A_2)$.
Similarly, $B_1 B_2 + B_3 B_4 = A_2 A_4 \sin(A_1 \widehat{A_2} A_3)$.
Let R be the radius of the circle. Now, by the Sine Rule applied to
$\triangle A_1 A_2 A_3$, we have $\frac{A_1 A_3}{\sin(A_1 \widehat{A_2} A_3)} = 2R$.
Also, by the Sine Rule applied to $\triangle A_0 A_1 A_2$, we have $\frac{A_0 A_2}{\sin(A_0 \widehat{A_1} A_2)} = 2R$.
The result now follows.

Problem 11.83. (This is Problem 10.20 on page 133.) *A chord ST of constant
length slides around a semi-circle with diameter AB.*
*Let M be the mid-point of ST and P be the foot of the perpendicular from
S to AB.*
*Prove that $\angle SPM$ has the same value, independent of the position of the
chord ST.*

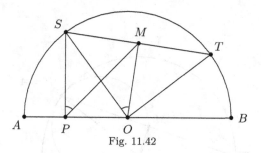

Fig. 11.42

Solution. To help, draw OS, OS and OT.

It is easy to show that $OM \perp ST$. Hence, $OPSM$ is a cyclic quadrilateral. Thus, $\angle SPM = \angle SOM = \frac{1}{2}\angle SOT$.

Since ST is a chord of fixed length, $\angle SOT$ is a fixed quantity. This proves the result.

Problem 11.84. (This is Problem 10.21 on page 133.) *Let Γ be a circle and P any fixed point.*

Consider the collection of all lines on P that intersect Γ. Suppose that a typical such line meets the circle at A and B.

Find the locus of the mid-point of AB.

Solution. Case I: P inside Γ.

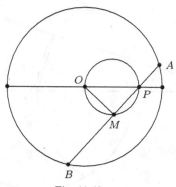

Fig. 11.43

The locus must include P (from the chord perpendicular to OP) and O (from the chord that includes OP).

Note that $OM \perp AB$, so that $\angle OMP = 90°$. Hence, M lies on a circle on OP as diameter.

Case II: P on Γ.

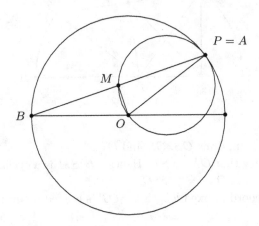

Fig. 11.44

Again, $\angle OMP = 90°$, so that M lies on a circle on OP as diameter.

Case III: P outside Γ.

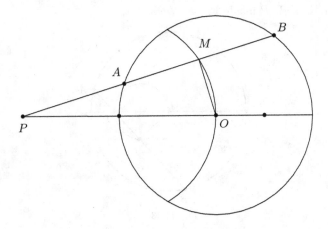

Fig. 11.45

Once again, we have $\angle OMP = 90°$, but we do not get the complete circle, since M is restricted to lie inside Γ.

Problem 11.85. (This is Problem 10.22 on page 133.) *Suppose that all the angles of △ABC are acute.*
Suppose that AD is an altitude.
Determine the points P and Q thus:

> *BP is parallel to AC and AP is perpendicular to AC;*
> *CQ is parallel to AB and AQ is perpendicular to AB.*

Show that AD bisects ∠PDQ.

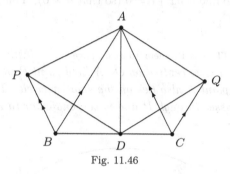

Fig. 11.46

Solution. By the "**Z**" Theorem, we have $\angle PBA = \angle BAC = \angle ACQ$.
Since $\angle APB = \angle AQC = 90°$, we have $\angle PAB = \angle QAC$.
Now quadrilaterals $APBD$ and $AQCD$ are cyclic (opposite angles $90°$).
Hence, $\angle PDB = \angle PAB = \angle QAC = \angle QDC$. Since $AD \perp BC$, we deduce that $\angle PDA = \angle QDA$; that is, AD bisects $\angle PDQ$.

Problem 11.86. (This is Problem 10.23 on page 133.) *Find the condition for the straight line $ax + by + c = 0$ to be tangent to the circle, centred at $(0,0)$ and of radius r.*

Solution. The circle is $x^2 + y^2 = r^2$. There are two cases.

Case I: $b \neq 0$. The line is then $y = \frac{-c - ax}{b}$. Substituting into the equation of the circle gives

$$x^2 + \left(\frac{-c - ax}{b} \right)^2 = r^2 \ ,$$

or

$$\left(1 + \frac{a^2}{b^2} \right) x^2 + \frac{2ac}{b^2} x + \left(\frac{c^2}{b^2} - r^2 \right) = 0 \ .$$

If the line is a tangent, this must be a perfect square; hence,

$$\left(\frac{2ac}{b^2}\right)^2 = 4\left(1 + \frac{a^2}{b^2}\right)\left(\frac{c^2}{b^2} - r^2\right) \quad ;$$
$$4a^2 c^a = 4\left(a^2 + b^2\right)\left(c^2 - b^2 r^2\right) \quad ;$$
$$r^2 = \frac{c^2}{a^2 + b^2} \quad .$$

Case II: $b = 0$. The line is $ax + c = 0$ (so that $a \neq 0$). Thus, $x = -\frac{c}{a} = \pm r$, or $r^2 = \frac{c^2}{a^2} = \frac{c^2}{a^2 + b^2}$.

Problem 11.87. (This is Problem 10.24 on page 133.) *In the diagram below, we have two fixed circles, centred on O. A fixed point P lies on the smaller circle. A variable point A also lies on the smaller circle. The chord BC of the larger circle passes through P and is perpendicular to AP.*

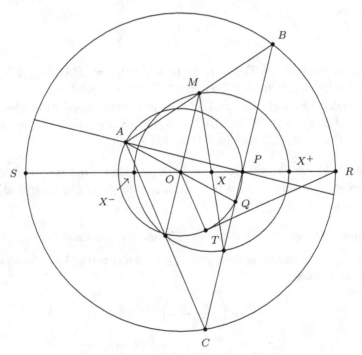

Fig. 11.47

a. Find the set of values of the expression $BC^2 + CA^2 + AB^2$.

b. Find the locus of the mid-point of AB (as A varies around the smaller circle).

Solution. First, let the radius of the larger circle be R and that of the smaller circle r.

a. From the intersecting Chords Theorem, we have
$$SP \cdot PR = (R - r)(R + r) = R^2 - r^2 = BP \cdot PC.$$

$$BC^2 + CA^2 + AB^2$$
$$= (BP + PC)^2 + CA^2 + AB^2$$
$$= BP^2 + 2(R^2 + r^2) + PC^2 + CA^2 + AB^2$$
$$= 2(R^2 + r^2) + BP^2 + PC^2 + (AP^2 + PC^2) + (AP^2 + BP^2)$$
$$= 2(R^2 + r^2) + 2(AP^2 + PC^2 + BP^2)$$
$$= 2(R^2 + r^2 + r^2 - PQ^2 + PC^2 + BP^2)$$
$$= 2(R^2 + PC^2 + BP^2 - PQ^2)$$
$$\qquad \text{since } PC = CQ + PQ = BP + PQ$$
$$= 2(R^2 + 2BP^2 + 2BP \cdot PQ)$$
$$= 2(R^2 + 2BP(BP + PQ)) = 2(R^2 + 2BP \cdot BQ)$$
$$= 2(R^2 + 2RT^2)$$
$$\qquad \text{using the Chord and Tangent Theorem}$$
$$= 2(R^2 + 2(R^2 - r^2)) = 6R^2 - 2r^2 = \text{constant.}$$

b. Suppose that the mid-point of OP is X, the mid-point of AB is M and the mid-point of AC is N. Then $OX = \frac{r}{2}$.
Consider the point X^+, lying between O and R, such that $OX^+ = \frac{r+R}{2}$, and the point X^-, lying between O and S, such that $OX^- = \frac{r-R}{2}$.
Then $X^- \cdot OX^+ = \frac{R^2 - r^2}{4}$.
Also, $MO \cdot ON = \frac{1}{2}BQ \cdot \frac{1}{2}QC = \frac{1}{4}BQ \cdot QC = \frac{1}{4}BP \cdot PC = \frac{1}{4}(R^2 - r^2)$.
Hence, $MO \cdot ON = X^- O \cdot OX^+$. Therefore, M, N, X^+ and X^- are concyclic. By symmetry, this is a circle on $X^+ X^-$ as diameter.

Problem 11.88. (This is Problem 10.25 on page 134.) *Given triangle ABC and points N_1, N_2 on AB, L_1, L_2 on BC, and M_1, M_2 on CA, respectively, such that*

$$AN_2 = N_2N_1 = N_1B = \frac{AB}{3} ,$$

$$BL_2 = L_2L_1 = L_1C = \frac{BC}{3} ,$$

$$CM_2 = M_2M_1 = M_1A = \frac{CA}{3} .$$

The points A_1, A_2, B_1, B_2, C_1 and C_2 are defined by

$$A_1 = M_1N_1 \cap M_2N_2 \quad A_2 = L_1N_1 \cap L_2M_2$$

$$B_1 = L_1N_1 \cap L_2N_2 \quad B_2 = L_1M_1 \cap N_2M_2$$

$$C_1 = M_1L_1 \cap M_2L_2 \quad C_2 = M_1N_1 \cap L_2N_2 .$$

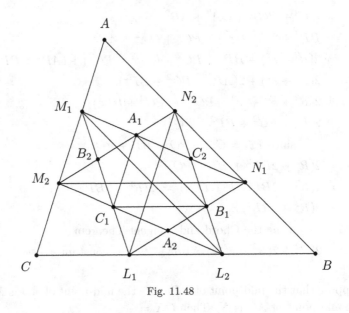

Fig. 11.48

(1) *Prove that* $L_1M_1 \parallel L_2N_2$.

(2) *Prove that* $A_1B_1 = A_2B_2$.

(3) *Calculate* $\dfrac{[A_1B_1C_1]}{[ABC]}$.

(4) *Calculate* $\dfrac{[A_1C_2B_1A_2C_1B_2]}{[ABC]}$.

Solution. (1) Note that $L_1N_2 \parallel AC$ since $\triangle ACB \approx \triangle N_2L_1B$.
Therefore, $\triangle CM_1L_1 \equiv \triangle L_1N_2L_2$, giving $\angle M_1L_1C = \angle N_2L_2L_2$.
Hence, $M_1L_1 \parallel L_2N_2$.

(2) From arguments similar to those above, $L_1N_1 \parallel M_2N_2$ and $L_2M_2 \parallel N_1M_1$.
Using similar arguments, we see that $M_1N_2 \parallel M_2N_1 \parallel B_2C_2 \parallel C_1B_1 \parallel CB$,
etc.

Consider $\triangle AA_1M_2$: we have $AA_1 \parallel M_1M_2$ and $AM_1 = M_1M_2$. There-
fore, $M_2B_2 = B_2A_1$ and $M_1B_2 = \frac{1}{2}AA_1$.
Thus, $M_2B_1 = B_2A_1 = A_1N_2$, plus two analogous statements.
Consider $\triangle M_1C_1C_2$: we have $M_1B_2 = B_2C_1$ and $M_1A_1 = A_1C_2$.
Therefore, $B_2A_1 \parallel C_1C_2$ and $B_2A_1 = \frac{1}{2}C_1C_2$.
Similarly, from $\triangle L_2C_1C_2$, we have $A_2B_1 = \frac{1}{2}C_1C_2$, and we are done.

(3) We have $A_1B_1 = N_2N_1 = \frac{1}{3}AB$.
Similarly, $B_1C_1 = \frac{1}{3}BC$ and $C_1A_1 = \frac{1}{3}CA$.
Hence, $\triangle A_1B_1C_1 \approx \triangle ABC$, with a similarity ratio of $\frac{1}{3}$. Therefore,
$[A_1B_1C_1] = \frac{1}{9}[ABC]$.

(4) $[A_2B_1C_1] = [L_1L_2A_2]$.
The base of $\triangle L_1L_2A_2$ is $\frac{1}{3}$ base of $\triangle ABC$.
The height of $\triangle L_1L_2A_2$ is $\frac{1}{3}$ height of $\triangle CM_2L_2$, which is $\frac{1}{3}$ $\left(\frac{1}{3}\right.$ height
of $\triangle ABC$), or $\frac{1}{9}$ of height $\triangle ABC$.
Therefore, $[A_2B_1C_1] = \frac{1}{27}[ABC]$.
Similarly $[A_1C_1B_2] = [A_1B_1C_2] = \frac{1}{27}[ABC]$.
Thus, $[A_1B_2C_1A_2B_1C_2] = \frac{1}{9}[ABC] + \frac{3}{27}[ABC] = \frac{2}{9}[ABC]$.

Problem 11.89. (This is Problem 10.26 on page 135.) *Let AB be a chord of a
circle centre O. Let OP be the radius of the circle that passes through the
mid-point of the chord AB. Let Q be any point on the circle. See Fig. 11.49.
Prove that PQ bisects the angle $\angle AQB$.*

Solution. We have acr $AP = $ **arc** BP; hence, $\angle AQP = \angle BQP$, and we
are done.

Problem 11.90. (This is Problem 10.27 on page 135.) *Given triangle ABC,
E any point on AC and F any point on AB. Suppose that BE and CF
intersect at D.
Prove that*

$$\frac{[ABC]}{[DBC]} = \frac{[AFE]}{[DFE]}.$$

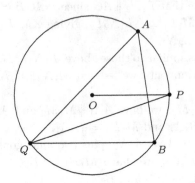

Fig. 11.49

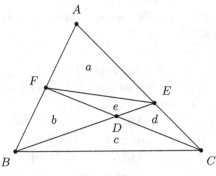

Fig. 11.50

Solution. Denote the areas of the various triangles by lower case letters as in the diagram above.

Since $\triangle DFE$ and $\triangle DCE$ have the same height, their areas are proportional to their bases. Therefore, $e : d = FD : DC$.

Similarly, $\triangle DFB$ and $\triangle DCB$ have the same height, their areas are proportional to their bases. Therefore, $b : c = FD : DC$.

Hence, $bd = ce$.

Similarly, considering $\triangle EAF$ and $\triangle ECF$, and $\triangle EAB$ and $\triangle ECA$, we deduce that $a(c + d) = (e + d)(a + e + b)$.

Similarly, considering $\triangle FAE$ and $\triangle FBE$, and $\triangle FAC$ and $\triangle FBC$, we deduce that $a(c + b) = (e + b)(a + e + d)$.

By expanding these, and using $bd = ce$, we have $ac = e(a + b + c + d + e)$, which proves the result.

This can also be done by Cartesian methods.

Let $B = (0,0)$, $C = (c,0)$, $A = (a,b)$.

Then $F = (1-\lambda)(0,0) + \lambda(a,b) = (\lambda a, \lambda b)$ and $E = (1-\mu)(c,0) + \mu(a,b) = (\mu a + (1-\mu)c, \mu b)$.

After finding the equations of BE and CF, we can calculate that D is

$$\left(\frac{\lambda(a\mu + c(1-\mu))}{\mu - \lambda - \lambda\mu}, \frac{b\lambda\mu}{\mu - \lambda - \lambda\mu} \right) .$$

We can now calculate the required areas using the standard determinant formula: this gives

$$[ABC] = \frac{bc}{2}, \qquad [BCD] = \left| \frac{bc\lambda\mu}{2(\mu - \lambda - \lambda\mu)} \right|,$$

and the ratio is $\left| \frac{\mu - \lambda - \lambda\mu}{\lambda\mu} \right|$.

Also,

$$[AFE] = \left| \frac{bc(1-\lambda)(1-\mu)}{2} \right|, \qquad [DFE] = \left| \frac{bc(1-\lambda)(1-\mu)}{2(\mu - \lambda - \lambda\mu)} \right|,$$

giving the same ratio as above.

Problem 11.91. (This is Problem 10.28 on page 136.) $ABCDE$, $DEFGH$ and $GHIJK$ are non-overlapping regular pentagons in a plane. Prove that $AJ = BK$.

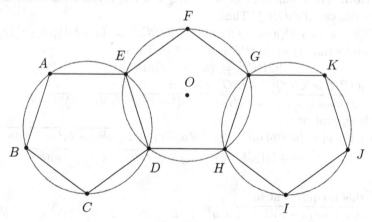

Fig. 11.51

Solution. The result follows by symmetry about O, the centre of pentagon $DEFGH$.

Problem 11.92. (This is Problem 10.29 on page 136.) *A, B and C are three points in order on a straight line such that $AB = 2a$ and $BC = 2b$. Semicircles are drawn on AB, BC and CA as diameters, all on the same side of the line ABC.*
Show that the radius of the circle drawn to touch all three semi-circles is given by $\frac{ab(a+b)}{a^2+ab+b^2}$.

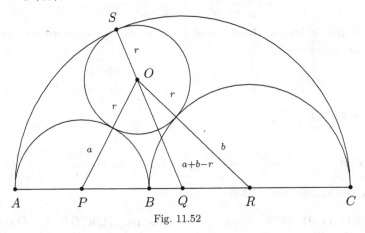

Fig. 11.52

Solution. The semiperimeter of $\triangle OPQ$ is $a+b$, which is the same as the semiperimeter of $\triangle ORQ$. Thus,

$$[OPQ]^2 = (a+b)(a)(b-r)(r) , \qquad [ORQ]^2 = (a+b)(b)(a-r)(r) .$$

The semiperimeter of $\triangle OPR$ is $a+b+r$. Thus,

$$[OPR]^2 = (a+b+r)(a)(b)(r) .$$

Since $[OPR] = [OPQ] + [ORQ]$, we have

$$\sqrt{rab(a+b+r)} = \sqrt{ra(a+b)(b-r)} + \sqrt{rb(a-r)(a+b)} .$$

This is equivalent to

$$ab(a+b+r) = (a+b)(a(b-r)+b(a-r)) + 2\sqrt{ab(a+b)^2(a-r)(b-r)}$$
$$= (a+b)(2ab-ar-br) + 2\sqrt{ab(a+b)^2(a-r)(b-r)} .$$

Now, this is equivalent to

$$2\sqrt{ab(a+b)^2(a-r)(b-r)} = a^2b + ab^2 + abr - 2a^2b - 2ab^2 + (a+b)^2r$$
$$= r(a^2 + 3ab + b^2) - ab(a+b) , \quad \text{or}$$

$$4ab(a+b)^2(a-r)(b-r) = \left((a+b)^2 + ab\right)^2 r^2$$
$$+ 2\left((a+b)^2 + ab\right)(ab)(a+b)$$
$$+ (ab)^2(a+b)^2 .$$

Solving yields

$$r = \frac{ab(a + b)}{a^2 + ab + b^2}$$

or

$$r = \frac{-3ab(a + b)}{a^2 + ab + b^2} \; .$$

The latter is clearly unacceptable.

Problem 11.93. (This is Problem 10.30 on page 136.) *A hexagon, two of whose sides are of length 2x, two are of length 2y and two are of length 2z, is inscribed in a circle. Prove that the radius ρ of this circle is given by the equation*

$$\rho^3 - (x^2 + y^2 + z^2)\rho = 2xyz \; .$$

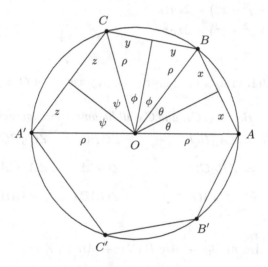

Fig. 11.53

Solution. We are not given the order of the sides. This does not matter, as is clear from the analysis. Thus, we suppose the hexagon to be $ABCA'C'B'$ with $AB = AB' = 2x$, $BC = B'C' = 2y$ and $CA' = C'A' = 2z$. Suppose that the radius of the circumcircle is ρ, that $\angle AOB = 2\theta$, that $\angle BOC = 2\phi$ and that $\angle COA' = 2\psi$. Drop perpendiculars from O to each of AB, BC and CA'.

Note that $\theta + \phi + \psi = \frac{\pi}{2}$ and that

$$\frac{x}{\rho} = \sin(\theta) \; , \qquad \frac{y}{\rho} = \sin(\phi) \; , \qquad \frac{z}{\rho} = \sin(\psi) \; .$$

Note that these results are independent of the configuration!
Now,

$$\sin(\theta) = \sin\left(\frac{\pi}{2} - \phi - \psi\right) = \cos(\phi + \psi)$$
$$= \cos(\phi)\cos(\psi) - \sin(\phi)\sin(\psi)$$
$$= \sqrt{1 - \sin^2(\phi)}\sqrt{1 - \sin^2(\psi)} - \sin(\phi)\sin(\psi) \; ;$$

$$\frac{x}{\rho} = \sqrt{1 - \frac{y^2}{\rho^2}}\sqrt{1 - \frac{z^2}{\rho^2}} - \frac{yz}{\rho^2} \; ;$$

$$x\rho + yz = \sqrt{(\rho^2 - y^2)(\rho^2 - z^2)} \; ;$$

$$x^2\rho^2 + 2xyz\rho + y^2z^2 = \rho^4 - \rho^2\left(y^2 + z^2\right) + y^2z^2 \; ;$$

$$\rho^4 - \rho^2\left(x^2 + y^2 + z^2\right) = 2xyz\rho \; ;$$

$$\rho^3 - \rho\left(x^2 + y^2 + z^2\right) = 2xyz \; ,$$

since $\rho \neq 0$.

Problem 11.94. (This is Problem 10.31 on page 136.) *Let O be the centre of* $\triangle ABC$.
Points A_B, A_C, B_C, B_A, C_B and C_A all lie on the circumcircle and satisfy;

$$\angle ABA_B = \angle A_B BO \; , \qquad\qquad \angle OBC_B = \angle C_B BC \; ,$$

$$\angle BCB_C = \angle B_C CO \; , \qquad\qquad \angle OCA_C = \angle A_C CA \; ,$$

$$\angle CAC_A = \angle C_A AO \; , \qquad\qquad \angle OAB_A = \angle B_A AB \; .$$

Prove that

$$\text{Arc } A_C B_C + \text{Arc } B_A C_A + \text{Arc } C_B A_B$$

$$= \text{Arc } A_B A_C + \text{Arc } B_C B_A + \text{Arc } C_A C_B \; .$$

Solution.

$$\left.\begin{array}{l}\text{Arc } B_A C_A = \frac{1}{2}\text{Arc } BC \\[4pt] \text{Arc } A_C B_C = \frac{1}{2}\text{Arc } AB \\[4pt] \text{Arc } C_B A_B = \frac{1}{2}\text{Arc } CA\end{array}\right\} \longrightarrow \quad \text{Arc } B_A C_A + \text{Arc } A_C B_C + \text{Arc } C_B A_B$$

is a semicircle.

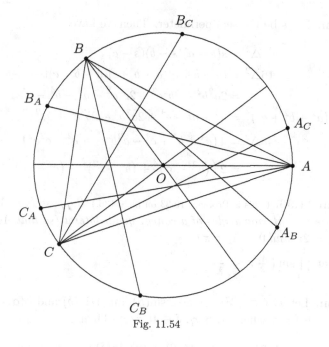

Fig. 11.54

$$\text{Arc } A_B A_C = \text{Arc } A_B A + \text{Arc } A A_C = \frac{1}{2}\left(\text{Arc } AB' + \text{Arc } AC'\right) \ ,$$

$$\text{Arc } B_C B_A = \text{Arc } B_C B + \text{Arc } B A_A = \frac{1}{2}\left(\text{Arc } BC' + \text{Arc } BA'\right) \ ,$$

$$\text{Arc } C_A C_A = \text{Arc } C_A C + \text{Arc } C C_B = \frac{1}{2}\left(\text{Arc } CA' + \text{Arc } AB'\right) \ .$$

Thus, **Arc** $A_B A_C$ + **Arc** $B_C B_A$ + **Arc** $C_A C_A$ is a semicircle.

Problem 11.95. (This is Problem 10.32 on page 136.) *If a, b, c are the sides of a triangle with area* Δ, *show that*

$$4\sqrt{3}\Delta \ \leq \ a^2 + b^2 + c^2 \ .$$

Solution. Let s be the semiperimeter. Then we have

$$\Delta^2 = s(s-a)(s-b)(s-c)$$
$$16\Delta^2 = (a+b+c)(a+b-c)(a-b+c)(-a+b+c)$$
$$= 2a^2b^2 + 2b^2c^2 + 2c^2a^2 - a^4 - b^4 - c^2 \ ;$$
$$\left(a^2+b^2+c^2\right)^2 = a^4 + b^4 + c^4 + 2a^2b^2 + 2b^2c^2 + ac^2a^2 \ ;$$
$$\left(a^2+b^2+c^2\right)^2 - 48\Delta^2 = 4\left(a^4 + b^4 + c^4 - a^2b^2 - b^2c^2 - c^2a^2\right)$$
$$= 2\left(\left(a^2-b^2\right)^2 + \left(b^2-c^2\right)^2 + \left(c^2-a^2\right)^2\right)^2 \geq 0 \ .$$

Problem 11.96. (This is Problem 10.33 on page 136.) *Suppose that A_i, $i = 1$, 2, 3, 4 are the interior angles of a convex quadrilateral (so that $A_1 + A_2 + A_3 + A_4 = 2\pi$ and $0 < A_i < \pi$).*
Show that $\prod\limits_{i=1}^{4} \sin\left(\frac{A_i}{2}\right) \leq \frac{1}{4}$.

Solution. Let $f(\alpha, \beta, \gamma, \delta) = \sin(\alpha)\sin(\beta)\sin(\gamma)\sin(\delta)$ and $\phi(\alpha, \beta, \gamma, \delta) = \alpha + \beta + \gamma + \delta = \pi$ where $\alpha, \beta, \gamma, \delta \in [0, \pi/2]$. Then

$$f_\alpha + \lambda\phi_\alpha = \cos\alpha\sin(\beta)\sin(\gamma)\sin(\delta) + \lambda = 0 \ ,$$
$$f_\beta + \lambda\phi_\beta = \sin\alpha\cos(\beta)\sin(\gamma)\sin(\delta) + \lambda = 0 \ ,$$
$$f_\gamma + \lambda\phi_\gamma = \sin\alpha\sin(\beta)\cos(\gamma)\sin(\delta) + \lambda = 0 \ ,$$
$$f_\delta + \lambda\phi_\delta = \sin\alpha\sin(\beta)\sin(\gamma)\cos(\delta) + \lambda = 0 \ .$$

Therefore, $\cos(\alpha)\sin(\beta) = \sin(\alpha)\cos(\beta)$, etc.;
that is, $\tan(\alpha) = \tan(\beta) = \tan(\gamma) = \tan(\delta)$.
Since $\alpha, \beta, \gamma, \delta \in [0, \pi/2]$, we deduce that $\alpha = \beta = \gamma = \delta = \pi/4$.
Thus, the extreme value is $f\left(\frac{\pi}{4}, \frac{\pi}{4}, \frac{\pi}{4}, \frac{\pi}{4}\right) = \frac{1}{\sqrt{2}} \cdot \frac{1}{\sqrt{2}} \cdot \frac{1}{\sqrt{2}} \cdot \frac{1}{\sqrt{2}} = \frac{1}{4}$.
Now, f is continuous and $f(0, 0, 0, \pi) = 0$. Thus, we have a maximum.

Problem 11.97. (This is Problem 10.34 on page 137.) *Suppose that A_i, $i = 1$, 2, 3, 4 are the interior angles of a convex quadrilateral (so that $A_1 + A_2 + A_3 + A_4 = 2\pi$ and $0 < A_i < \pi$).*
Show that $\prod\limits_{i=1}^{4} \cos\left(\frac{A_i}{2}\right) \leq \frac{1}{4}$.

Solution. The proof is similar to the previous one.

Problem 11.98. (This is Problem 10.35 on page 137.) *Suppose that A_i, $i = 1$, 2, 3 are the interior angles of an acute angled triangle.*

Show that $\prod_{i=1}^{3} \tan(A_i) \geq 3\sqrt{3}$.

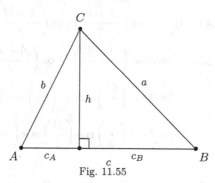

Fig. 11.55

Solution. Note that $\tan(A) = \frac{h}{c_A}$ and $\tan(B) = \frac{h}{c_B}$.

Let $f(A, B, C) = \tan(A)\tan(B)\tan(C)$ and $\phi(A, B, C) = A + B + C = \pi$, where $0 \leq A, b, C \leq \pi/2$. Therefore,

$$f_A + \lambda\phi_A = \sec^2(A)\tan(B)\tan(C) + \lambda = 0 ,$$
$$f_B + \lambda\phi_B = \sec^2(B)\tan(A)\tan(C) + \lambda = 0 ,$$
$$f_C + \lambda\phi_C = \sec^2(C)\tan(A)\tan(B) + \lambda = 0 .$$

Thus,

$$\sec^2(A)\tan(B) = \tan(A)\sec^2(B) , \text{ or } \sin(A)\cos(A) = \sin(B)\cos(B) , \text{ etc.}$$

Therefore, $A = B = C = \pi/3$ and $f\left(\frac{\pi}{3}, \frac{\pi}{3}, \frac{\pi}{3}\right) = 3\sqrt{3}$.

Since f is continuous and $f\left(0, 0, \frac{\pi}{2}\right) = +\infty$, we have a minimum.

Problem 11.99. (This is Problem 10.36 on page 137.)

Suppose that $\angle A + \angle B + \angle C = 180°$. Prove that

$$\sin(A) + \sin(B) + \sin(C) = 4\cos\left(\frac{A}{2}\right)\cos\left(\frac{B}{2}\right)\cos\left(\frac{C}{2}\right) .$$

Solution.

$$\sin(A) + \sin(B) = 2\sin\left(\frac{A+B}{2}\right)\cos\left(\frac{A-B}{2}\right) \quad ;$$

$$\sin(C) = \sin(\pi - A - B) = 2\sin\left(\frac{A+B}{2}\right)\cos\left(\frac{A+B}{2}\right) \quad ;$$

$$\sin(A) + \sin(B) + \sin(C) = 2\sin\left(\frac{A+B}{2}\right)\left(\cos\left(\frac{A+B}{2}\right) + \cos\left(\frac{A-B}{2}\right)\right)$$

$$= 2\cos\left(\frac{C}{2}\right)\cdot 2\cos\left(\frac{A}{2}\right)\cos\left(\frac{B}{2}\right)$$

$$= 4\cos\left(\frac{A}{2}\right)\cos\left(\frac{B}{2}\right)\cos\left(\frac{C}{2}\right) \quad .$$

Problem 11.100. *Suppose that* $\angle A + \angle B + \angle C = 180°$. *Prove that*

$$\begin{vmatrix} 1 & 1 & 1 \\ \cos(A) & \cos(B) & \cos(C) \\ \tan\left(\frac{A}{2}\right) & \tan\left(\frac{B}{2}\right) & \tan\left(\frac{C}{2}\right) \end{vmatrix} = 0 \quad .$$

Solution. Now,

$$\begin{vmatrix} 1 & 1 & 1 \\ \cos(A) & \cos(B) & \cos(C) \\ \tan\left(\frac{A}{2}\right) & \tan\left(\frac{B}{2}\right) & \tan\left(\frac{C}{2}\right) \end{vmatrix} = \begin{vmatrix} 1 & 0 & 0 \\ \cos(A) & \cos(B) - \cos(A) & \cos(C) - \cos(A) \\ \tan\left(\frac{A}{2}\right) & \tan\left(\frac{B}{2}\right) - \tan\left(\frac{A}{2}\right) & \tan\left(\frac{C}{2}\right) - \tan\left(\frac{A}{2}\right) \end{vmatrix}$$

and this expands to

$$(\cos(B) - \cos(A))\left(\tan\left(\frac{C}{2}\right) - \tan\left(\frac{A}{2}\right)\right)$$
$$- (\cos(C) - \cos(A))\left(\tan\left(\frac{B}{2}\right) - \tan\left(\frac{A}{2}\right)\right) \quad .$$

If this is zero, then it is sufficient to prove that

$$\frac{\cos(B) - \cos(A)}{\cos(C) - \cos(A)} = \frac{\left(\tan\left(\frac{B}{2}\right) - \tan\left(\frac{A}{2}\right)\right)}{\left(\tan\left(\frac{C}{2}\right) - \tan\left(\frac{A}{2}\right)\right)} \quad .$$

Now,

$$\begin{aligned}
\frac{\cos(B) - \cos(A)}{\cos(C) - \cos(A)} &= \frac{2\sin\left(\frac{B-A}{2}\right)\sin\left(\frac{B+A}{2}\right)}{2\sin\left(\frac{C-A}{2}\right)\sin\left(\frac{C+A}{2}\right)} \\
&= \frac{\sin\left(\frac{2B+C}{2} - \frac{\pi}{2}\right)\sin\left(\frac{\pi}{2} - \frac{C}{2}\right)}{\sin\left(\frac{B+2C}{2} - \frac{\pi}{2}\right)\sin\left(\frac{\pi}{2} - \frac{B}{2}\right)} \\
&= \frac{\cos\left(\frac{2B+C}{2}\right)\cos\left(\frac{C}{2}\right)}{\cos\left(\frac{B+2C}{2}\right)\cos\left(\frac{B}{2}\right)} \\
&= \frac{\cos\left(\frac{C}{2}\right)}{\cos\left(\frac{B}{2}\right)} \cdot \frac{\sin\left(\frac{B+C}{2}\right)\sin\left(\frac{B}{2}\right) - \cos\left(\frac{B+C}{2}\right)\cos\left(\frac{B}{2}\right)}{\sin\left(\frac{B+C}{2}\right)\sin\left(\frac{C}{2}\right) - \cos\left(\frac{B+C}{2}\right)\cos\left(\frac{C}{2}\right)} \\
&= \frac{\frac{\sin\left(\frac{B}{2}\right)}{\cos\left(\frac{B}{2}\right)} - \frac{\cos\left(\frac{B+C}{2}\right)}{\sin\left(\frac{B+C}{2}\right)}}{\frac{\sin\left(\frac{C}{2}\right)}{\cos\left(\frac{C}{2}\right)} - \frac{\cos\left(\frac{B+C}{2}\right)}{\sin\left(\frac{B+C}{2}\right)}} \\
&= \frac{\tan\left(\frac{B}{2}\right) - \cot\left(\frac{B+C}{2}\right)}{\tan\left(\frac{C}{2}\right) - \cot\left(\frac{B+C}{2}\right)} \\
&= \frac{\tan\left(\frac{B}{2}\right) - \tan\left(\frac{A}{2}\right)}{\tan\left(\frac{C}{2}\right) - \tan\left(\frac{A}{2}\right)} \ .
\end{aligned}$$

Problem 11.101. (This is Problem 10.27 on page 135.) *Suppose that* $\angle A + \angle B + \angle C = 180°$. *Prove that*

$$\frac{1}{\sin(A/2)} + \frac{1}{\sin(B/2)} + \frac{1}{\sin(C/2)} \geq 6 \ .$$

Solution. Let $F(A, B, C) = \frac{1}{\sin(A/2)} + \frac{1}{\sin(B/2)} + \frac{1}{\sin(C/2)}$ and $\phi(A, B, C) = A + B + C - \pi$. Then

$$F_A + \lambda\phi_A = \frac{\cos(A/2)}{2\sin^2(A/2)} + \lambda \ ,$$

$$F_B + \lambda\phi_B = \frac{\cos(B/2)}{2\sin^2(B/2)} + \lambda \ ,$$

$$F_C + \lambda\phi_C = \frac{\cos(C/2)}{2\sin^2(C/2)} + \lambda \ .$$

Hence, $\sin(A/2)\tan(A/2) = \sin(B/2)\tan(B/2) = \sin(C/2)\tan(C/2)$. Since $\sin(x)\tan(x)$ is increasing on $[0, \pi/2)$, we obtain $A = B = C = \pi/3$. Since $F(A, B, C)$ is continuous and tends to infinity as any of A, B, C tends to zero, we have a minimum of $F(\pi/3, \pi/3, \pi/3) = 6$.

Problem 11.102. (This is Problem 10.38 on page 137.) *Suppose that R and r represent the circumradius and inradius of $\triangle ABC$. Prove that*

$$\frac{2R}{r} = \frac{\sin(A) + \sin(B) + \sin(C)}{\sin(A)\sin(B)\sin(C)} \geq 4 \;.$$

Prove also that

$$\frac{\sin(A) + \sin(B) + \sin(C)}{\sin(A)\sin(B)\sin(C)} = 4$$

if and only if $A = B = C = 60°$.

Solution. We have to find the minimum of
$$f(A,B,C) = \frac{1}{\sin(A)\sin(B)} + \frac{1}{\sin(B)\sin(C)} + \frac{1}{\sin(C)\sin(A)}.$$
Using the Sine Rule in the form $\frac{a}{\sin(A)} = 2R$, etc., we have

$$f(A,B,C) = 4R^2 \left(\frac{1}{ab} + \frac{1}{bc} + \frac{1}{ca} \right)$$
$$= 4R^2 \left(\frac{a+b+c}{abc} \right) = \frac{8R^2 s}{abc}$$
$$= \frac{2R \cdot 4Rs}{4Rsr} = \frac{2R}{s} \;,$$

where we have used the known result: $abc = 4Rs$.

The remainder of the proof follows the lines of the previous example.

Problem 11.103. (This is Problem 10.39 on page 137.) *Consider a trapezoid $ABCD$ with $AB \parallel CD$. Let P be the point of intersection of the diagonals AC and BD.*

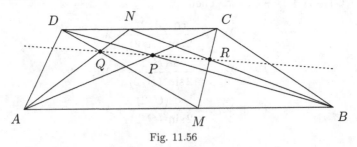

Fig. 11.56

Choose any point M on AB and produce MP to meet DC at N. Let AN and MD meet at Q. Let BN and MC meet at R. Prove that P, Q and R lie on a line.

Solution. Normally, such a result would be proved by the techniques of Projective Geometry. But that is beyond the scope of this book. Therefore, we proceed using Cartesian methods.

Let the points be $A = (0,0)$, $B = (b,0)$, $C = (c,1)$ and $D = (d,1)$. Let $M = (\lambda b, 0)$.

Then the equation of AC is $y = \frac{x}{c}$ and the equation of BD is $y = \frac{b-x}{b-d}$.

Solving together gives $P = \left(\frac{bc}{b-d+c}, \frac{b}{b-d+c} \right)$.

The equation of MP is $y = (x - \lambda b) \left(\frac{\frac{b}{b-d+c}}{\frac{bc}{b-d+c} - \lambda b} \right)$, and this leads to

$N = (c + \lambda(d - c), 1)$.

We may now obtain the equations of NA and DM, and solve them together to find Q, and similarly, find R.

Then we can find the equation of QR and check that P lies on it.

The details are let to the reader.

Problem 11.104. (This is Problem 10.40 on page 137.) *In an acute angled triangle ABC with orthocentre H and circumcentre O, show that*

$$\angle HAO = |\angle ABC - \angle ACB| \ .$$

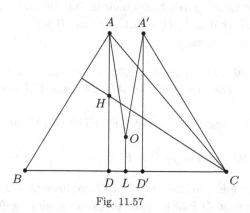

Fig. 11.57

Solution. Let D be the foot of the altitude AD, and let A' and D' be the reflections of A and D in OL, respectively.

Then $\angle A'CB = \angle ABC$ by symmetry, and $AA'BC$ is a cyclic quadrilateral with centre O (since $OA = OA' = OB = OC$). Hence, $\angle AOA' = 2\angle ACA'$. Since $AD \parallel OL \parallel A'D'$, we have $\angle HAO = \frac{1}{2}\angle AOA'$.

Since $\angle ACA' = \angle A'CB - \angle ACB = \angle ABC - \angle ACB$, the result follows.

Problem 11.105. (This is Problem 10.41 on page 138.) *Let AD, BE and CF be the internal bisectors of the angles of ABC, with D, E and F on BC, CA and AB, respectively.*
If quadrilateral BCEF is cyclic, show that ABC is isosceles.

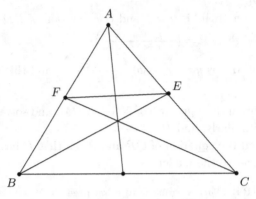

Fig. 11.58

Solution. Since we are given that quadrilateral $BCEF$ is cyclic, we have $\angle FBE = \angle FCE$; that is $\frac{1}{2}\angle B = \frac{1}{2}\angle C$, so that $B = C$.
Hence, $\triangle ABC$ is isosceles.

Problem 11.106. (This is Problem 10.42 on page 138.) *Triangle ABC has circumcentre O. The mid-points of AB, BC and CA are F, D and E, respectively.*
Produce DO, EO and FO to meet EF, FD and DE at D', E' and F', respectively.
Show that O is the incentre of D'E'F'.

Solution. Since $FE \parallel BC$ and $OD \perp BC$, we have $OD' \perp FE$, etc.
Thus, quadrilateral $OD'FE'$ is cyclic (opposite angles are $90circ$).
Therefore, $\angle OE'D' = \angle OFD'$.
Since quadrilateral $OFAE$ is cyclic, $(\angle OFD' =) \angle OFE = \angle OAE$.
Since $\triangle OCA$ is isosceles $(OC = OC = R)$, $\angle OAE = \angle ECO$.
Since quadrilateral $OECD$ is cyclic, $\angle ECO = \angle ODE$.
Since quadrilateral $OE'DF'$ is cyclic, $(\angle ODE =) \angle ODF' = \angle OE'F'$.
Thus, OE' bisects $\angle D'E'F'$, etc.
Therefore, O is the incentre of $\triangle D'E'F'$.

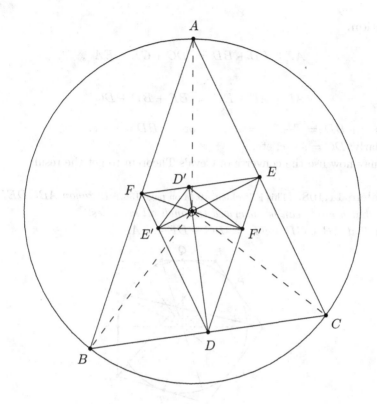

Fig. 11.59

Problem 11.107. (This is Problem 10.43 on page 138.) *In triangle ABC, let D be the point on BC such that AB + BD = AC + CD. Similarly, let E and F be on AC and AB, respectively, such that BA + AE = BC + CE and CA + AF = CB + BF.*
Show that AD, BE and CF are concurrent.

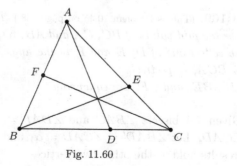

Fig. 11.60

Solution.

$$AF + FB + BD = DC + CE + EA \; ,$$

$$AF + AE + EC = BF + BD + DC \; .$$

Thus, $c + BD = \frac{a+b+c}{2} = s$, $\quad$ or $\quad$ $BD = s - c$.
Similarly, $DC = s - c$, etc.
We may now use the converse of Ceva's Theorem to get the result.

Problem 11.108. (This is Problem 10.44 on page 138.) *Hexagon ABCDEF is such that a circle can be inscribed to touch all six sides. Show that $AB + CD + EF = BC + DE + FA$.*

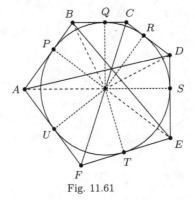

Fig. 11.61

Solution. Since AU and AP are tangents, we have $AP = AU$, and similar results at each vertex of the hexagon. The result now follows.
CHALLENGE: prove that AD, BE and CF are concurrent.

Problem 11.109. (This is Problem 10.45 on page 138.) *In triangle ABC, let D, E and F be the mid-points of BC, CA and AB, respectively. Let D', E' and F' be the reflections of D, E and F in the angle bisectors of $\angle CAB$, $\angle ABC$ and $\angle BCA$, respectively. Show that AD', BE' and CF' are concurrent.*

Solution. Given IA bisects $\angle BAC$ and $\angle DAI = \angle D'AI$. Therefore, $\angle BAD' = \angle CAD$. Let $\angle BAD' = \angle CAD = \alpha_1$ and $\angle D'AI = \angle DAI = \alpha_2$. Similar results hold at the other two vertices.

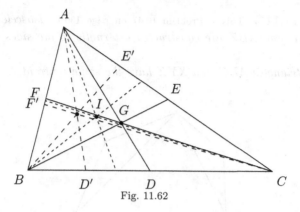

Fig. 11.62

The concurrency of the medians, AD, BE, CE, implies that

$$\prod_{\text{cyclic}} \frac{\sin(\alpha_1)}{\sin(\alpha_1+2\alpha_2)} = 1.$$

The concurrency of AD', BE', CF', requires

$$\prod_{\text{cyclic}} \frac{\sin(\alpha_1+2\alpha_2)}{\sin(\alpha_1)} = 1,$$ which we know to be true.

Problem 11.110. (This is Problem 10.46 on page 138.) *In triangle ABC, the points D, E and F lie on BC, CA and AB, respectively, and are such that $BC = 3BD$, $CA = 3CE$ and $AB = 3AF$.*
Show that triangles ABC and DEF have the same centroid.

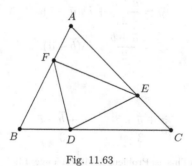

Fig. 11.63

Solution. Let $\overline{a}$ be the vector representing the point A, etc.
Then $\overline{g} = \frac{\overline{a}+\overline{b}+\overline{c}}{3}$ and $\overline{d} = \frac{\overline{c}}{3} + \frac{2\overline{b}}{3}$.
Thus, $\overline{g} = \frac{1}{3}\left(\left(\frac{\overline{c}}{3} + \frac{2\overline{b}}{3}\right) + \left(\frac{\overline{a}}{3} + \frac{2\overline{c}}{3}\right) + \left(\frac{\overline{b}}{3} + \frac{2\overline{a}}{3}\right)\right) = \frac{\overline{a}+\overline{b}+\overline{c}}{3}.$
Therefore, G is the centroid of $\triangle DEF$.

Problem 11.111. (This is Problem 10.47 on page 139.)   *Isosceles triangles BCX, CAY and ABZ are constructed externally to the sides of triangle ABC.*
Show that triangles ABC and XYZ have the same centroid.

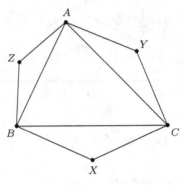

Fig. 11.64

Solution. Using the notations of the previous example for the complex numbers representing each point:

$$\overline{x} = \frac{\overline{b} + \overline{c}}{2} + i\lambda(\overline{c} - \overline{b}) \ ,$$

$$\overline{y} = \frac{\overline{c} + \overline{a}}{2} + i\lambda(\overline{a} - \overline{c}) \ ,$$

$$\overline{z} = \frac{\overline{a} + \overline{b}}{2} + i\lambda(\overline{b} - \overline{a}) \ .$$

Therefore,

$$\frac{\overline{x} + \overline{y} + \overline{z}}{3} = \frac{\overline{a} + \overline{b} + \overline{c}}{3} \ .$$

Problem 11.112. (This is Problem 10.48 on page 139.)   *For triangle ABC, denote the lengths for the altitudes from A, B and C by h_a, h_b and h_c, respectively.*
Extend the altitudes at A, B and C through the opposite sides to A', B' and C', respectively, where $AA' = \frac{k}{h_a}$, $BB' = \frac{k}{h_b}$ and $CC' = \frac{k}{h_c}$, k being a positive constant.
Prove that triangles ABC and A'B'C' have the same centroid.

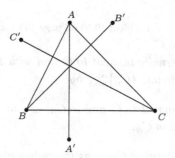

Fig. 11.65

Solution. Use the complex variable techniques of the previous example.

Problem 11.113. (This is Problem 10.49 on page 139.) *In* $\triangle ABC$*, let O be the centre. Draw perpendiculars from O to each side: $OD \perp BC$, $OE \perp CA$ and $OF \perp AB$.*
Extend each to D', E' and F', respectively, so that $OD = DD'$, $OE = EE'$ and $OF = FF'$.
Prove that $\triangle D'E'F'$ is congruent to $\triangle ABD$.

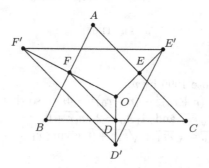

Fig. 11.66

Solution. $AC \parallel FD$ because F and D are the mid-points of AB and CB.
$FD \parallel F'D'$ because F and D are the mid-points of OF' and OD'.
Therefore, $AC \parallel F'D'$, and further, $\triangle ABC \approx \triangle D'E'F'$.
Now, because of F and D being mid-points of line segments, we have
$[D'E'F'] = 4[DEF] = [ABC]$. Thus, $\triangle ABC \equiv \triangle DEF$.

Problem 11.114. (This is Problem 10.50 on page 139.) *Let C be a fixed circle and T a tangent line to C.*
Let C_0 be a circle, externally tangent to C and with T as a tangent line. Define a sequence of circles: $\{C_n\}_{n=1}^{\infty}$ by:

a. C_n *is externally tangent to C and to C_{n-1}; and*
b. T *is a tangent line to C_n.*

Find the locus of the centre of C_{1999} as the radius of C_0 is allowed to vary in size.

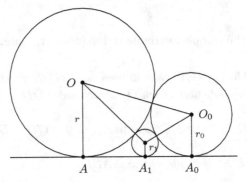

Fig. 11.67

Solution. We denote 1999 by n.
We have $(AA_0)^2 = (r + r_0)^2 - (r - r_0)^2 = 4rr_0$, so that $AA_0 = 2\sqrt{rr_0}$.
Similarly, $AA_1 = 2\sqrt{rr_1}$ and $A_1A_0 = 2\sqrt{r_1r + 0}$.
Thus, we have $\sqrt{rr_0} = \sqrt{rr_1} + \sqrt{r_1r + 0}$, giving $r_1 = \dfrac{rr_0}{\left(\sqrt{r}+\sqrt{r_0}\right)^2}$.
Similarly,

$$r_2 = \frac{rr_1}{\left(\sqrt{r} + \sqrt{r_1}\right)^2}$$
$$= \frac{rr_0}{\left(\sqrt{r} + 2\sqrt{r_0}\right)^2} \ ,$$
$$\vdots$$
$$r_n = \frac{rr_0}{\left(\sqrt{r} + n\sqrt{r_0}\right)^2} \ .$$

Also, $AA_n = 2\sqrt{rr_n} = \dfrac{2r\sqrt{r_0}}{\left(\sqrt{r}+n\sqrt{r_0}\right)} \ .$

If we take A to be the origin, we than have the coordinates of the centre O_n as $\left(\frac{2r\sqrt{r_0}}{(\sqrt{r}+n\sqrt{r_0})}, \frac{2r\sqrt{r_0}}{(\sqrt{r}+n\sqrt{r_0})} \right)$.

Therefore, the locus lies on the parabola $x^2 = 4ry$, which is independent of r_0.

Problem 11.115. (This is Problem 10.51 on page 139.) *By using an appropriate placement of a triangle in the Cartesian coordinate system, prove the following:*

(a) The medians of a triangle are concurrent.
(b) The altitudes of a triangle are concurrent.

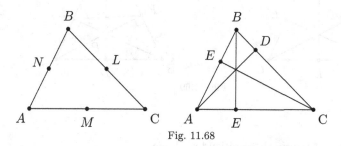

Fig. 11.68

Solution. Let $A = (0,0)$, $B = (2a, 2b)$ and $C = (2,0)$.
(a) Then $L = (1 + a, b)$, $M = (1, 0)$ and $N = (a, b)$.
The equation of AL is $y = \frac{b}{a+1} x$.
The equation of BM is $y = \frac{2b}{2a-1} x - \frac{2b}{2a-1}$.
The equation of CN is $y = \frac{b}{a-2} x - \frac{2b}{a-2}$.
The lines AL and BM meet at $\left(\frac{2(a+1)}{3}, \frac{2b}{3} \right)$.
It is easy to check that this lies on CN.
As a bonus, we see that the intersection point is $\frac{1}{3}$ up BM, and by symmetry, $\frac{1}{3}$ up each median.

(b) The equation of AD is $y = \frac{1-a}{b} x$.
The equation of BE is $x = 2a$.
The equation of CE is $y = \frac{-a}{b} x + \frac{2a}{b}$.
The lines AD and CE meet at $\left(2a, \frac{2a(1-a)}{b} \right)$.
It is easy to check that this lies on CN.

Problem 11.116. (This is Problem 10.52 on page 139.) *Prove that the altitudes of a triangle are concurrent:*

a. *using Ceva's Theorem;*
b. *using only properties of triangles and circles, but NOT Ceva's Theorem or Menelaus' Theorem.*

a. 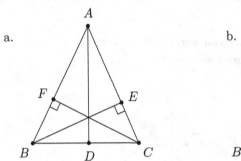 b.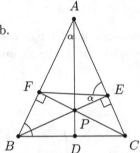

Fig. 11.69

Solution. a.   Assume three altitudes. Then

$$\frac{BD}{DC} = \frac{\frac{BD}{AB} \cdot AB}{\frac{DC}{AC} \cdot AC} = \frac{\cos(B) \cdot AB}{\cos(C) \cdot AC} \ .$$

Thus,

$$\frac{BD}{DC} \cdot \frac{CE}{EA} \cdot \frac{AF}{FB} = \frac{\cos(B)}{\cos(C)} \cdot \frac{AB}{AC} \cdot \frac{\cos(C)}{\cos(A)} \cdot \frac{BC}{BA} \cdot \frac{\cos(A)}{\cos(B)} \cdot \frac{CA}{CB} = 1 \ .$$

b.   Altitudes BE and CF meet at P. Extend AP to meet BC at D.
Since $\angle AFP = \angle AEP = 90°$, quadrilateral $AFPE$ is cyclic.
Hence, $\angle FAP = \angle FEP$ (angles in the same segment — marked with dots).
Since $\angle BFC = \angle BEC = 90°$, quadrilateral $BFEC$ is cyclic.
Hence, $\angle AEF = \angle FBC$ (exterior angle property — marked with arcs).
But, $\angle AEF + \angle FEP = 90°$. Hence, $\angle FAP + \angle FBC = 90°$;
that is, $\angle BAD + \angle ABD = 90°$, so that $\angle ADB = 90°$.
Hence, AD is an altitude.

Problem 11.117. (This is Problem 10.53 on page 140.) *Suppose that* Γ *is a fixed circle and AB a fixed chord of that circle.*
Let C be any other point on Γ. *Let G be the centroid of* $\triangle ABC$.
Find the locus of G as C varies around the circle Γ.

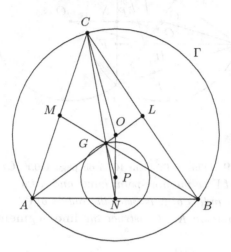

Fig. 11.70

Solution. We know that $GN = \frac{1}{3}CN$. Construct P such that $PN = \frac{1}{3}ON$.
Then $GP \parallel OC$ and $GP = \frac{1}{3}OC = \frac{R}{3}$, where R is the radius of Γ.
Hence, the locus of G is a circle, centre P of radius $\frac{R}{3}$.

Problem 11.118. (This is Problem 10.54 on page 140.) *Suppose that* Γ *is a fixed circle and AB a fixed chord of that circle.*
Let C be any other point on Γ. *Let H be the orthocentre of* $\triangle ABC$.
Find the locus of H as C varies around the circle Γ.

Solution. Let N be the mid-point of AB, let G be the centroid, let P be the point of ON such that $OP = 2PN$. Then $CO \parallel GP$ and $OC = 3PG$ (as in previous example).
Note that OGH is the Euler line and $HG = 2GO$. Find Q on OP extended such that $QP = 2PO$. Then $QH \parallel PG$ and $QH = 3PG$.
Therefore, $QH = OC = R$ (where R is the radius of Γ). This is a constant.
Thus, the locus of H is a circle centre Q and radius R.

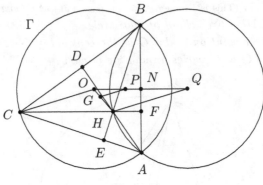

Fig. 11.71

Problem 11.119. (This is Problem 10.55 on page 140.) *Given $\triangle ABC$ with orthocentre H, let Γ be the nine-point circle and Λ the circumcircle.*
(a) Show that the radius of Λ is twice the radius of Γ.
*Let P be any point on Λ. Construct the **line segment** PH and let it intersect Γ at Q.*
(b) Prove that $HQ = QP$.

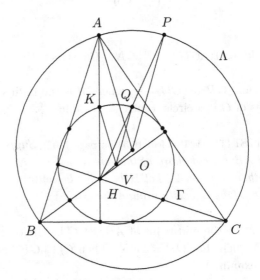

Fig. 11.72

Solution. (a) The centre of the nine-point circle, V, is the mid-point of OH. Also, K is the mid-point of AH. Hence, $VK \parallel OA$ and $VK = \frac{1}{2}OA$. Thus, the radius of the nine-point circle is one half of the radius of the circumcircle.

(b) Since $VQ = \frac{1}{2}OP$ and $HV = \frac{1}{2}HO$, we have that Q is the mid-point of PH.

Problem 11.120. (This is Problem 10.56 on page 140.) *In $\triangle ABC$, the points C_A, C_1 and C_B lie on the line segment AB such that $AC_A = C_A C_1 = C_1 C_B = C_B B$. Similarly, the points B_A, B_1 and B_C lie on the line segment AC such that $AB_A = B_A B_1 = B_1 B_C = B_C C$, and the points A_B, A_1 and A_C lie on the line segment BC such that $BA_B = A_B A_1 = A_1 A_C = A_C C$. Show that $\triangle ABC$, $\triangle A_B B_C C_A$ and $\triangle A_C C_B B_A$ all have the same centroid.*

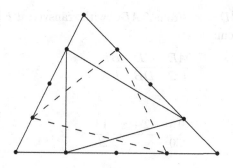

Fig. 11.73

Solution. This can be done using vectors.
Let $A = \overline{a}$, $B = \overline{b}$ and $C = \overline{c}$.
Then $A_B = \frac{1}{4}\overline{b} + \frac{3}{4}\overline{a}$, etc.

Problem 11.121. (This is Problem 10.57 on page 140.) *Triangle ABC has sides $AB = 39cm$, $BC = 42cm$ and $CA = 45cm$. Points E and F lie on the line segments AC and AB, respectively, so that*

$$AE = \frac{AC}{3} \quad and \quad BF = \frac{BA}{3},$$

respectively. The line EF meets the line BC at D.
Calculate the length of the line segment BD.

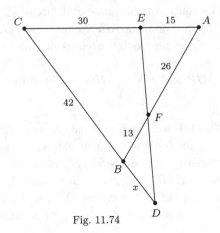

Fig. 11.74

Solution. Let $BD = x$. From $\triangle ABC$ with transversal EFD, we get from Menelaus' Theorem:

$$\frac{AE}{EC} \cdot \frac{CD}{DB} \cdot \frac{BF}{FA} = -1 .$$

Thus,

$$\frac{15}{30} \cdot \frac{42 + x}{x} \cdot \frac{13}{26} = 1 ,$$

so that $x = 14$.

Problem 11.122. (This is Problem 10.58 on page 140.)   *Triangle ABC is inscribed in a circle* Γ. *The internal bisectors of the angles* $\angle CAB$, $\angle ABC$ *and* $\angle BCA$ *meet* Γ *again at U, V and W, respectively.*
Perpendiculars for U, V and W meet the sides BC, CA and AB, respectively, at L, M and N, respectively.
Prove that AL, BM and CN are concurrent.

Solution. We have $\angle BAU = \angle CAU$ and quadrilateral $ABUC$ is cyclic. Hence, $\angle BCU = \angle BAU = \angle CAU = \angle CBU$.
Also, $\angle BLU = \angle CLU = 90°$. Thus, we have $\triangle BLU \equiv \triangle CLU$, and thus, $BL = CL$; that is, L is the mid-point of BC.
Similarly, M is the mid-point of CA and N is the mid-point of AB. Hence, AL, BM and CN are concurrent.

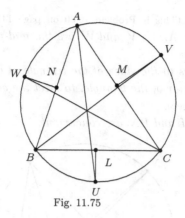

Fig. 11.75

Problem 11.123. (This is Problem 10.59 on page 141.) *Triangle XYZ has sides XY = 24, YZ = 16 and ZX = 20. Points E and F lie on the line segments XZ and XY, respectively, so that*

$$XE = \frac{XZ}{4} \quad and \quad YF = \frac{YX}{4},$$

respectively. The line EF meets the line YZ at D.

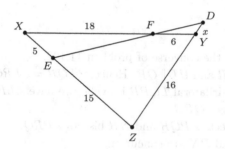

Fig. 11.76

Calculate the length of the line segment DY.

Solution. Using Menelaus' Theorem for $\triangle XYZ$ and transversal EFD, we have

$$\frac{YF}{FX} \cdot \frac{XE}{EZ} \cdot \frac{ZD}{DY} = -1 .$$

Thus,

$$\frac{6}{18} \cdot \frac{5}{15} \cdot \frac{16 + x}{x} = 1 ,$$

giving $x = 2$.

Problem 11.124. (This is Problem 10.60 on page 141.) *Triangle PQR is inscribed in a circle Λ. U, V and W are the mid-points of QR, RP and PQ, respectively.*
Draw perpendiculars to the sides of the triangle at U, V and W outwards away from the interior of the triangle, to meet the circle Λ at the points L, M and N, respectively.
Prove that PL, QM and RN are concurrent.

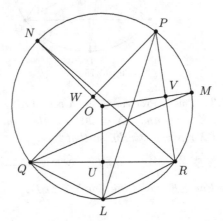

Fig. 11.77

Solution. This is the converse of problem 11.122.
We have $QU = UR$ and $UL \parallel QR$. Hence, $\angle LQR = \angle LRQ$.
Further, since quadrilateral $LQPR$ is cyclic, we have $\angle LPR = \angle LPQ$.
That is, PL bisects $\angle QPR$.
Similarly, MQ bisects $\angle PQR$ and NR bisects $\angle PRQ$.
Thus, PL, QM and RN are concurrent.

Problem 11.125. (This is Problem 10.61 on page 141.) *Given quadrilateral ABCD and a transverse line PQRS, where P lies on AB, Q lies on BD, R lies on AC and S lies on DC (none being coincident with any vertex of the quadrilateral).*
Show that
$$\frac{AP}{PB} \cdot \frac{DS}{SC} = \frac{AR}{RC} \cdot \frac{DQ}{QB} .$$

Solution. Let AD and PS meet at Z.
Consider $\triangle ABD$ with transversal ZPQ. We have
$$\frac{DZ}{ZA} \cdot \frac{AP}{PB} \cdot BQQD = -1 .$$

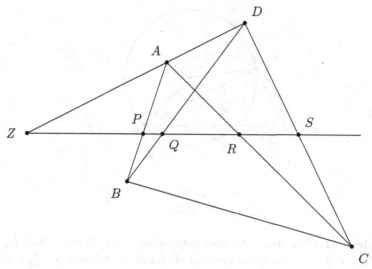

Fig. 11.78

Consider $\triangle ACD$ with transversal ZRS. We have

$$\frac{DZ}{ZA} \cdot \frac{AR}{RC} \cdot \frac{CS}{SD} = -1 \ .$$

Therefore, $\frac{AP}{PB} \cdot \frac{BQ}{QD} = \frac{AR}{RC} \cdot \frac{CS}{SD}$, and the result follows.

Problem 11.126. (This is Problem 10.62 on page 142.)

   a. Given triangle $\triangle ABC$, let D, E and F be the mid-points of BC, CA and AB, respectively. The circle, centre E and radius EC, intersects the circle, centre D and radius DC, at C and P. Show that P lies on AB.

   b. Similarly, find Q on BC and R on CA. Show that AQ, BR and CP are concurrent.

   c. Determine what this point of concurrency is.

Solution. a. Note that $\angle APC = 90°$ (angle in a semicircle) and $\angle BPC = 90°$ (angle in a semicircle). Hence, P lies on AB.

  b. Note that CP is an altitude of $\triangle ABC$. Similarly, AQ and BR are altitudes. Thus, they are concurrent.

  c. The orthocentre.

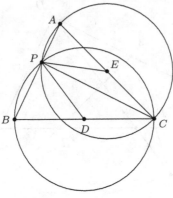

Fig. 11.79

Problem 11.127. (This is Problem 10.63 on page 142.) *Suppose that* I_A, I_B *and* I_C *are the centres of the escribed circles of the triangle ABC, opposite to A, B and C, respectively. Show that the centre of the inscribed circle of the triangle ABC is the orthocentre of the triangle* $I_A I_B I_C$.

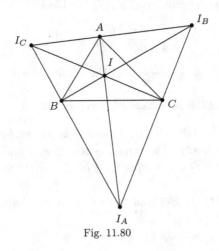

Fig. 11.80

Solution. Because we have bisectors, we have $I_C A I_B \perp A I I_A$, etc.

Problem 11.128. (This is Problem 10.64 on page 142.) *The inscribed circle of triangle ABC touches the sides BC, CA and AB in X, Y and Z, respectively. I is the centre of the circle.*
If XI intersects ZY in L, prove that AL is a median of triangle ABC.

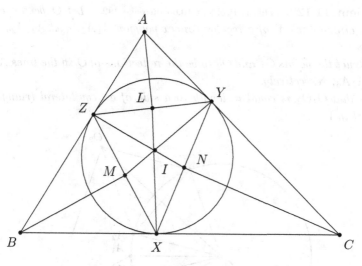

Fig. 11.81

Solution. Let $\angle ZAL = A_1$, $\angle YAL = A_2$, $\angle XBM = B_1$, $\angle ZBM = B_2$,
$\angle YCN = C_1$ and $\angle XCN = C_2$.
Note that $AZ = AY$, $BX = BZ$ and $CX = CY$.
Let $\angle AZL = \angle AYZ = \alpha$.
Note that $\alpha = \frac{\pi}{2} - \frac{A}{2}$.
Let $\angle BZM = \angle BXM = \beta$.
Note that $\beta = \frac{\pi}{2} - \frac{B}{2}$.
Using the Sine Rule, we have

$$\frac{\sin(\alpha)}{AL} = \frac{\sin(A_1)}{ZL}, \qquad \frac{\sin(\alpha)}{AL} = \frac{\sin(A_2)}{LY}.$$

Hence, $\frac{\sin(A_1)}{\sin(A_2)} = \frac{ZL}{LY}$.
Similarly, $\frac{\sin(B_1)}{\sin(B_2)} = \frac{XM}{MZ}$ and $\frac{\sin(C_1)}{\sin(C_2)} = \frac{YN}{NX}$.
But, XL, YM and ZN are concurrent.
Therefore, $\frac{ZL}{LY} \cdot \frac{YN}{NX} \cdot \frac{XM}{MZ} = 1$.
Thus, $\frac{\sin(A_1)}{\sin(A_2)} \cdot \frac{\sin(B_1)}{\sin(B_2)} \cdot \frac{\sin(C_1)}{\sin(C_2)} = 1$.
But, this is the angle form of Ceva's Theorem!
Therefore, AL, BM and CN are concurrent.

Problem 11.129. (This is Problem 10.65 on page 142.) *Let O be the centre of the circumcircle Γ of a regular convex polygon $A_1 A_2 A_3 \ldots A_{25} A_{26}$ of 26 sides.*
Construct the points O_1 and O_2 to be the reflections of O in the lines $A_{25}A_1$ and $A_2 A_6$, respectively.
Prove that $O_1 O_2$ is equal in length to a side of an equilateral triangle inscribed in Γ.

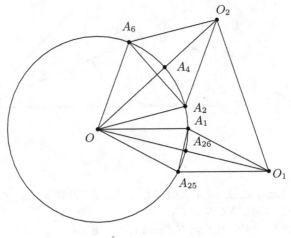

Fig. 11.82

Solution. Denote the radius of the circle by r. Note that $OO_1 = 2r\cos(\pi/13)$ and $OO_2 = 2r\cos(2\pi/13)$. Also, $\angle O_1 O O_2 = 4\pi/13$.
Using the Cosine Rule, we have

$$
\begin{aligned}
O_1 O_2^2 &= OO_1^2 + OO_2^2 - 2OO_1 \cdot OO_2 \cos(4pi/13) \\
&= 4r^2 \left(\cos^2(\pi/13) + \cos^2(2\pi/13) - 2\cos(\pi/13)\cos(2\pi/13)\cos(4\pi/13) \right) \\
&= 3r^2
\end{aligned}
$$

if and only if

$$
\cos^2(\pi/13) + \cos^2(2\pi/13) - 2\cos(\pi/13)\cos(2\pi/13)\cos(4\pi/13) = \frac{3}{4}
$$
$$
= \cos^2(\pi/6) \ .
$$

Now, computer algebra (DERIVE) tells me that this is correct, but I do not yet have a direct proof.

Problem 11.130. (This is Problem 10.66 on page 142.) *Consider any non-equilateral triangle with circumcentre O, centroid G, in-centre I and orthocentre H. Let V be the centre of the nine point circle.*
Let P be the foot of the perpendicular from I to the Euler line OGVH.
Prove that P lies between G and H.

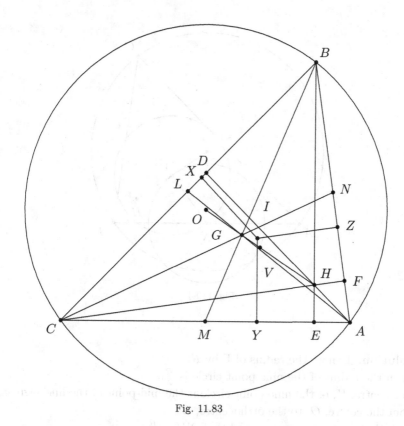

Fig. 11.83

Solution. Since X lies between D and L, etc., and V lies between G and H on the Euler Line, we see that V lies between AL and AD, etc., and thus, V lies in a convex polygon with GH as a diagonal.
Hence, P lies between G and H.

Problem 11.131. (This is Problem 10.67 on page 142.) *Suppose that* Γ *is a fixed circle and* AB *a fixed chord of that circle.*
Let C *be any other point on* Γ. *Let* J *be the centre of the nine point circle of* $\triangle ABC$.
Find the locus of J *as* C *varies around the circle* Γ.

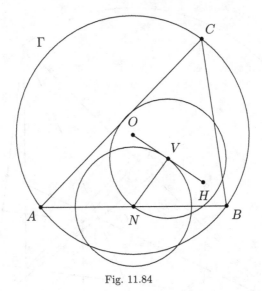

Fig. 11.84

Solution. Denote the radius of Γ by R.
Then the radius of the nine point circle is $\frac{R}{2}$.
The centre, V, of the nine point circle is the mid-point of the line segment from the centre, O, to the orthocentre, H.
Note that N is a fixed point and that $NV = \frac{R}{2}$.
Thus, V lies on a circle, radius $\frac{R}{2}$, centre N.

Problem 11.132. (This is Problem 10.68 on page 143.) *In* $\triangle ABC$, *let* D, E *and* F *be points in the interiors of the line segments* BC, CA *and* AB, *respectively. Suppose that* AD, BE *and* CF *are concurrent at* P. *Let* X, Y *and* Z *be the mid-points of* PA, PB *and* PC, *respectively.*
Show that the circumradius of $\triangle XYZ$ *is half the circumradius of* $\triangle ABC$.

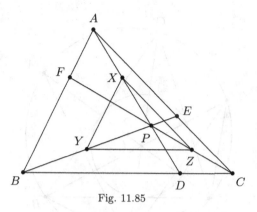

Fig. 11.85

Solution. Because of the mid-points, we see that $\triangle ABC$ and $\triangle XYZ$ are similar, with a similarity ration of $\frac{1}{2}$.

Therefore, $[ABC] = 4[XYZ]$.

Now, denoting the circumradius of $\triangle ABC$ by R and that of $\triangle XYZ$ by ρ, and using $[ABC] = \frac{abc}{4R}$, we get

$$\rho = \frac{\frac{1}{2}a\,\frac{1}{2}b\,\frac{1}{2}c}{4[XYZ]} = \frac{1}{2}\frac{abc}{4[ABC]} = \frac{1}{2}R.$$

Problem 11.133. (This is Problem 10.69 on page 143.) *Suppose that the sides of right $\triangle ABC$ are all integers. Prove that the inradius is an integer.*

Solution. The sides of an integer right triangle are given by Pythagorean triples, generated by

$$z(x^2 - y^2) \ , \qquad 2xyz \ , \qquad z(x^2 + y^2) \ ,$$

where $x > y$, z are positive integers.

The area of $\triangle ABC$ is $[ABC] = \frac{1}{2}\left(2xyz \times z(x^2 - y^2)\right) = xyz^2(x+y)(x-y)$ and the semi-perimeter is

$$s = \frac{1}{2}\left(z(x^2 + y^2) + 2xyz + z(x^2 + y^2)\right) = xz(x + y) \ .$$

Hence, the inradius is $r = \frac{[ABC]}{s} = yz(x - y)$, which is an integer.

Note that if the given triangle is placed on lattice points, the the incentre is at a lattice point.

Problem 11.134. (This is Problem 10.70 on page 143.) *Suppose that Γ is a fixed circle and AB a fixed chord of that circle.*

Let C be any other point on Γ. Let I be the incentre of $\triangle ABC$.

Find the locus of I as C varies around the circle Γ.

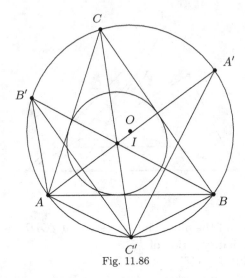

Fig. 11.86

Solution. Note that the internal bisector of the $\angle ACB$ meets the circumcircle again at the point C', where $OC' \perp AB$. Similarly for B' and C'.

Note that $\angle AB'A' = \angle ACA' = \angle BAA' = \angle BB'A'$. Similarly, $\angle AA'B' = \angle CA'B'$. Therefore, $\triangle AA'B' \equiv \triangle IA'B'$, giving $AA' = IA'$.

But AA' is a fixed length. Therefore, IA' is a fixed length.

Thus, the locus of I consists of the two arcs of the circles, on A and B, with centre at the point on the circumcircle on the chord through the mid-point of AB, perpendicular to AB, on the opposite side of AB from C, inside the circumcircle.

Problem 11.135. (This is Problem 10.71 on page 143.) *Suppose that Γ is a fixed circle and AB a fixed chord of that circle.*

Let C be any other point on Γ. Let I_A, I_B and I_C be the centres of the escribed circles of $\triangle ABC$.

Find the locus of I_A, I_B and I_C as C varies around the circle Γ.

Solution. The previous problem gives that $IC' = BC'$, so that I_C lies on the same circle that I lies on.

By similar arguments to the previous problem, we have $\angle AI_B B = \angle C'B'B = \frac{1}{2}\angle AB'B = \frac{1}{2}\angle ACB$, and $\angle AI_A B = \angle AA'C' = \frac{1}{2}\angle AA'B = \frac{1}{2}\angle ACB$.

Thus, $ABI_A I_B$ is a cyclic quadrilateral, so that I_A and I_B lie on the same circle.

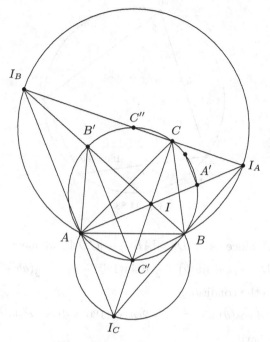

Fig. 11.87

Since the given circle is the nine-point circle of $\triangle I_A I_B I_C$ (it passes through the feet of the altitudes), we have that C'' is the mid-point of $I_A I_B$, and since $\angle I_A A I_B$ is a right angle, it follows that C'' is the centre of circle $ABI_A I_B$. Note that C'' is on the perpendicular bisector of AB (and is therefore, diametrically opposite to C').

When C is on the same arc as C', then I_A and I_B lie on the lower circular arc, and I_C lies on the upper circular arc.

Problem 11.136. (This is Problem 10.72 on page 143.) *Obtain the following for* **cyclic quadrilaterals** *with sides $AB = a$, $BC = b$, $CD = c$, $DA = d$ and diagonals $AC = e$, $BD = f$, and where $s = \dfrac{a+b+c+d}{2}$ and R is the radius of the circle:*

(1) $[ABCD] = \sqrt{(s-a)(s-b)(s-c)(s-d)}$;

(2) $e = \sqrt{\dfrac{(ac+bd)(bc+ad)}{ab+cd}}$;

(3) $R = \dfrac{1}{4}\sqrt{\dfrac{(ab+cd)(ac+bd)(bc+ad)}{(s-a)(s-b)(s-c)(s-d)}}$.

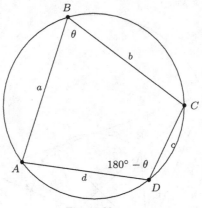

Fig. 11.88

Solution. (1) Since $[ABCD] = [ABC] + [CDA]$, we have

$$[ABCD] = \tfrac{1}{2}ab\sin(\theta) + \tfrac{1}{2}cd\sin(180° - \theta) = \tfrac{1}{2}(ab + cd)\sin(\theta) ,$$

subject to the condition:

$$a^2 + b^2 - 2ab\cos(\theta) = c^2 + d^2 - 2cd\cos(180° - \theta) = c^2 + d^2 + 2cd\cos(\theta) .$$

Thus, we have

$$(ab + cd)\sin(\theta) = 2[ABCD] ,$$
$$(ab + cd)\cos(\theta) = \tfrac{1}{2}\left(a^2 + b^2 - c^2 - d^2\right) .$$

Squaring and adding gives

$$(ab + cd)^2 = 4[ABCD]^2 + \tfrac{1}{4}\left(a^2 + b^2 - c^2 - d^2\right)^2 ,$$

or

$$\begin{aligned}
16[ABCD]^2 &= 4(ab + cd)^2 - \left(a^2 + b^2 - c^2 - d^2\right)^2 \\
&= \left(2(ab + cd) + \left(a^2 + b^2 - c^2 - d^2\right)\right) \\
&\quad \times \left(2(ab + cd) - \left(a^2 + b^2 - c^2 - d^2\right)\right) \\
&= \left(a^2 + 2ab + b^2 - c^2 + 2cd - d^2\right) \\
&\quad \times \left(-a^2 + 2ab - b^2 + c^2 + 2cd + d^2\right) \\
&= \left((a + b)^2 - (c - d)^2\right)\left((c + d)^2 - (a - b)^2\right) \\
&= (a + b + c - d)(a + b - c + d) \\
&\quad \times (a - b + c + d)(-a + b + c + d) \\
&= (2s - 2d)(2s - 2c)(2s - 2b)(2s - 2a) \\
&= 16(s - a)(s - b)(s - c)(s - d) .
\end{aligned}$$

(2) We make use of Ptolemy's Theorem, which gives us that $ac + bd = ef$.
Let R be the radius of the circle.
We have $[ABC] = \frac{abe}{4R}$ and $[CDA] = \frac{cde}{4R}$.
Thus, $[ABCD] = \frac{(ab+cd)e}{4R}$. Similarly, $[ABCD] = \frac{(ad+bd)f}{4R}$.
Therefore,

$$[ABCD]^2 = \frac{(ab+cd)(ad+bc)ef}{16R^2} = \frac{(ab+cd)(ad+bc)(ac+bd)}{16R^2} \ .$$

Now use the first result, and we get the desired formula for R.
(3) We have

$$e^2 = \frac{\frac{4R[ABCD]}{f}}{\frac{4R[ABCD]}{e}} \, ef = \frac{bc+ad}{ab+cd} \, ef \ .$$

Now use Ptolemy's Theorem, and we get the result.

Problem 11.137. (This is Problem 10.73 on page 143.) *Suppose that quadrilateral ABCD has an incircle of radius r.*
If $AB = a$, $BC = b$, $CD = c$, $DA = d$, and $s = \dfrac{a+b+c+d}{2}$, prove that

(1) $a + c = b + d$;
(2) $[ABCD] = rs$.

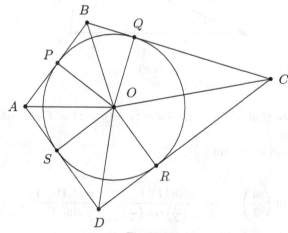

Fig. 11.89

Solution. (1) Note that $OP = OQ = OR = OS = r$. Let $AP = AS = \alpha$, $BP = BQ = \beta$, $CQ = CR = \gamma$ and $DR = DS = \delta$. Then

$$a = \alpha + \beta \ , \quad c = \gamma + \delta \ ,$$
$$b = \beta + \gamma \ , \quad d = \delta + \alpha \ .$$

We immediately have $a + c = b + d$.

(2) Note that $s = \frac{1}{2}(a + b + c + d) = \alpha + \beta + \gamma + \delta$.
Since $OP \perp AB$, etc., we have

$$[OPAS] = \alpha r \ , \quad [OQBP] = \beta r \ , \quad [ORCQ] = \gamma r \ , \quad [OSDR] = \delta r \ .$$

Hence, $[ABCD] = r(\alpha + \beta + \gamma + \delta) = rs$.

Problem 11.138. *Show that* $r = \dfrac{2R\sin^2(A)}{1+\cos(A)}$.

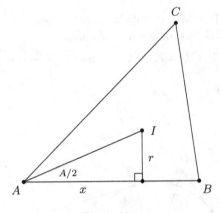

Fig. 11.90

Solution. Note that $s - a = (x + y + z) - (y + z) = x = r \cot\left(\frac{A}{2}\right)$, and that $a = 2R\sin(A)$.
Thus, $s = 2R\sin(A) + r \cot\left(\frac{A}{2}\right)$.
Since

$$\cot\left(\frac{A}{2}\right) = \frac{2\cos^2\left(\frac{A}{2}\right)}{2\sin\left(\frac{A}{2}\right)\cos\left(\frac{A}{2}\right)} = \frac{\cos(A) + 1}{\sin(A)} \ ,$$

the result follows.

Problem 11.139. *Show that*

$$s \tan\left(\frac{A}{2}\right) + 2R\cos(A) = 2R + r \ .$$

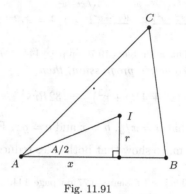

Fig. 11.91

Solution. Note that

$$\frac{r}{s-a} = \frac{r}{x} = \tan\left(\frac{A}{2}\right)$$

$$= \frac{2\sin^2\left(\frac{A}{2}\right)}{2\sin\left(\frac{A}{2}\right)\cos\left(\frac{A}{2}\right)}$$

$$= \frac{1 - \cos(A)}{\sin(A)} \ .$$

Hence,

$$\frac{ra}{s-a} = \frac{a}{\sin(A)}(1 - \cos(A)) = 2R(1 - \cos(A)) \ .$$

Since

$$\frac{ra}{s-a} = r\left(\frac{s}{s-a} - 1\right) = s\frac{r}{x} - r \ ,$$

the result follows.

Problem 11.140. (This is Problem 10.75 on page 144.) *Show that, if the sides of a triangle are in arithmetic progression, then*

$$2s^2 = 9Rr - 18r^2 \ .$$

Solution. Suppose that $a = 2x - y$, $b = 2x$ and $c = 2x + y$. Thus, $s = 3x$. Now, $\Delta^2 = s(s-a)(s-b)(s-c) = 3x(2x+y)x(2x-y) = 3x^2(x^2-y^2)$. Also, $r = \frac{\Delta}{s}$, so that $r^2 = \frac{3x^2(x^2-y^2)}{9x^2} = \frac{x^2-y^2}{3}$, and $r = \frac{\sqrt{x^2-y^2}}{\sqrt{3}}$. Further, $R = \frac{abc}{4\Delta} = \frac{2x(4x^2-y^2)}{\sqrt{3}x\sqrt{x^2-y^2}} = \frac{2(4x^2-y^2)}{\sqrt{3}\sqrt{x^2-y^2}}$. Thus, $Rr - 2r^2 = \frac{2(4x^2-y^2)}{3} = \frac{2(x^2-y^2)}{3} = 2x^2 = \frac{2}{9}s^2$. The result follows.

Problem 11.141. (This is Problem 10.76 on page 144.) *Show that, if the sides of a triangle are in geometric progression, then*

$$\left(s^2 + 4Rr + r^2\right)^3 = 32Rrs^4 .$$

Solution. Suppose that $a = x/g$, $b = x$ and $c = gx$. Follow the techniques of the previous problem, to show that both sides reduce to $x^6\left(g + 1 + \frac{1}{g}\right)^3$.

Problem 11.142. (This is Problem 10.77 on page 144.) *Prove the following equalities about the angles of $\triangle ABC$:*

(1) $\sin(A) + \sin(B) + \sin(C) = \dfrac{s}{R}$,

(2) $\cos(A) + \cos(B) + \cos(C) = \dfrac{R+r}{R}$,

(3) $\cos(A)\cos(B) + \cos(B)\cos(C) + \cos(C)\cos(A) = \dfrac{r^2 + s^2 - 4r^2}{4R^2}$,

(4) $\cos(A)\cos(B)\cos(C) = \dfrac{s^2 - (2R+r)^2}{4R^2}$,

(5) $\tan(A) + \tan(B) + \tan(C) = \dfrac{2rs}{s^2 - 4R^2 - 4Rr - r^2}$

$$= \dfrac{2rs}{s^2 - (2R+r)^2} ,$$

(6) $\tan(A/2) + \tan(B/2) + \tan(C/2) = \dfrac{4R+r}{s}$.

Solution. (1) From the Sine Rule, $\sin(A) = \frac{a}{2R}$. The others are similar. The result follows.

(2) From the Cosine Rule, $\cos(A) = \frac{b^2+c^2-a^2}{2bc}$. The others are similar. Adding gives

$$\frac{a(b^2 + c^2 - a^2) + b(c^2 + a^2 - b^2) + c(a^2 + b^2 - c^2)}{2abc} .$$

Subtract 1, from this. The result factors to

$$\frac{(a+b-c)(a-b+c)(-a+b+c)}{2abc} .$$

Multiply numerator and denominator by $a + b + c$: we get

$$\frac{(a+b+c)(a+b-c)(a-b+c)(-a+b+c)}{2abc(a+b+c)} = \frac{16[ABC]^2}{2(4[ABC]R)2s}$$

$$= \frac{[ABC]}{sR} = \frac{rs}{sR} = \frac{r}{R},$$

and the result follows.

(3) Follow the technique of part 2.

(4) Follow the technique of part 2.

(5) Follow the technique of parts 1 and 2.

(6) Note that $\tan(A/2) = \frac{r}{x}$, etc., that $abc = 4Rrs$ and that $xyz = \frac{\Delta^2}{s} = r^2 s$. Thus,

$$\sum \tan\left(\frac{A}{2}\right) = r\left(\frac{1}{x} + \frac{1}{y} + \frac{1}{z}\right)$$

$$= \frac{r}{xyz}(xy + yz + zx)$$

$$= \frac{1}{rs}(xy + yz + zx)$$

$$= \frac{1}{rs}\left(x(y + z - (x + y + z)) + y(z - (x + y + z))\right.$$
$$\left. -z(x + y + z) + (x + y + z)^2\right)$$

$$= \frac{1}{rs}\left(x(y + z - s) + y(z - s) - sz + s^2\right)$$

$$= \frac{1}{rs^2}\left(sx(y + z - s) + sy(z - s) - s^2 z + ss\right)$$

$$= \frac{1}{rs^2}\left(s^3 - s^2(x + y + z) + s(xy + yz + zx) - sr^2 + sr^2\right)$$

$$= \frac{1}{rs^2}\left(s^3 - s^2(x + y + z) + s(xy + yz + zx) - xyz + sr^2\right)$$

$$= \frac{1}{rs^2}\left((s - x)(s - y)(s - z) + rs^2\right)$$

$$= \frac{1}{rs^2}\left(abc + r^2 s\right)$$

$$= \frac{1}{rs^2}\left(4Rrs + rs^2\right) = \frac{4R + r}{s}.$$

Problem 11.143. (This is Problem 10.78 on page 144.) *H is the orthocentre of ABC. Given the points A, B and H, find a straight edge and compass construction to find the point C.*

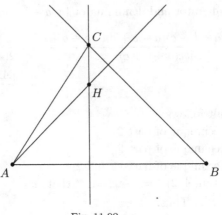

Fig. 11.92

Solution. Through H, draw a perpendicular to AB.

Join AH and extend it.

Through B, draw a perpendicular to AH, to meet the perpendicular through H to AB at the point C.

Then the required triangle is $\triangle ABC$.

Proof.

CH is perpendicular to AB, and AH is perpendicular to BC. Hence, BH is perpendicular to AC.

Problem 11.144. (This is Problem 10.79 on page 144.) *Show how to construct triangle with sides equal to and parallel to the medians of a given triangle.*

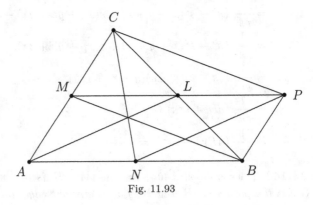

Fig. 11.93

Solution. Draw $NP \parallel AL$ with $NP = AL$. Join CP. Then the required triangle is $\triangle CNP$.

Proof.

Join PL and LM. Then $ALPN$ is a parallelogram, so that $LP \parallel AN$ and $LP = AN$.

But, $ML \parallel AN$ and $ML = AN$. Hence, MLP is a straight line and l is the mid-point of MP.

Also, L is the mid-point of BC.

Hence, $CPBM$ is a parallelogram, and thus, $CP = MB$.

Problem 11.145. *Give a purely geometric proof that*

$$\tan^{-1}\frac{1}{2} + \tan^{-1}\frac{1}{3} = \frac{\pi}{4}.$$

Solution. We start with right triangle ABC, where $\angle ABC = \frac{\pi}{2}$ and $\angle BAC = \tan^{-1}\frac{1}{2}$. WLOG, let $AB = 6$ and $BC = 3$. Then $AC = 3\sqrt{5}$. Draw square $ACDE$ with B in the interior of the square. Extend CB to meet DE at F. Let AC and CF intersect at G.

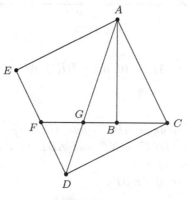

Fig. 11.94

First, note that $\angle FCD = \tan^{-1}\frac{1}{2}$. This implies that F is the mid-point of DE.

Note that triangles ACG and DFG are similar. Since $DF = \frac{1}{2}AC$, we have that $DG = \frac{1}{2}AG$. Since $AD = 3\sqrt{10}$, it follows that $AG = 2\sqrt{10}$ and further, $BG = 2$.

Hence, $\angle BAG = \tan^{-1}\frac{1}{3}$. But $\angle DAC = \frac{\pi}{4}$.

We can now conclude that $\tan^{-1}\frac{1}{2} + \tan^{-1}\frac{1}{3} = \frac{\pi}{4}$.

Problem 11.146. *On the planet, Nobis, bisection of line segments by straight edge and compasses is forbidden.*
You are commanded by King Youclid to start with line segments of lengths 16 and 12, and, with straight edge and compasses, construct a line segment of length 7.
Can you obey the king's command?

Solution. Simply draw a diagram with rectangle $ABCD$.

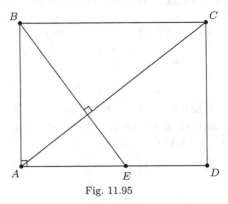

Fig. 11.95

$AB = DC = 12$, $BC = AD = 16$, draw $BE \perp AC$.
Therefore, $ED = 7$.

Problem 11.147. *Rectangle $ABCD$ has $AB = \frac{BC}{2}$. Outside the triangle, draw $\triangle DCF$, where $\angle DFC = 30°$ and ADF is a straight line segment. Let E be the mid-point of AD.*
Determine the measure of $\angle EBF$.

Solution. We extend the rectangle to a square, the half equilateral triangle to an equilateral triangle, and add another equilateral triangle, labelled as shown:
It is easy to see that $\angle BCF = \angle IGF = 150°$.
Thus, $\angle CBF = \angle CFB = \angle GFI = 15°$.
Hence, $\angle BFJ = 60°$.
Since $\angle BJF = 90°$, we have $\angle JBF = \angle EBF = 30°$.

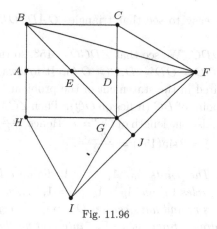

Fig. 11.96

Problem 11.148. *A square of side length s is inscribed symmetrically inside a sector of a circle with radius of length r and central angle of* 60°, *such that two vertices lie on the straight sides of the sector and two vertices lie on the circular arc of the sector.*
Determine the exact value of $\frac{s}{r}$.

Solution. Start with a semicircle, centre O and diameter POV. Draw radii $OP, OB, OQ, OE, OS, OF, OT, OH, OU, OI, OV$ in order such that the angle between each successive pair is 15°.
Points A, D, G and J are on OP, OR, OT and OV, respectively, such that $BA \parallel QO \parallel CD$, $ED \parallel SO \parallel FG$, and $HG \parallel UO \parallel IJ$.

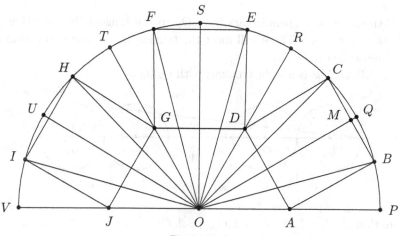

Fig. 11.97

By symmetry, it is easy to see that triangles OAD, ODG and OGJ are equilateral triangles.

Consider triangle ODC. We see that $\angle DCO = 15°$, so that triangle ODC is isosceles, giving $CD = OD = DA$ ($= OA$). It follows that $ABCD$ is a square as was required in the statement of the problem.

Let M be the mid-point of BC (it lies on OQ). Then $\angle CMO = 90°$ and the length of CM is $\frac{s}{2}$. Also the length of $OC = r$. Hence, $\frac{CM}{CO} = \frac{s}{2r} = \sin(15°)$. It now follows that $\frac{s}{r} = 2\sin(15°) = \frac{\sqrt{6}-\sqrt{2}}{2}$.

Problem 11.149. *The points A_0, A_1, $\ldots$, A_n lie on a line, ordered from left to right. The circles Γ_k on $A_{k-1}A_k$ ($k = 1, \ldots, n$) as diameters all have the same radius r, and have centres O_k, respectively.*

A tangent line is drawn from A_0 to Γ_n, intersecting the circle Γ_k at the points B_k and C_k. Here, $B_1 = A_0$, $B_n = C_n = P$, the point of tangency on Γ_n, and the point B_k is to the left of the point C_k.

Determine the length of the line segment B_kC_k, where this tangent intersects the circle Γ_k.

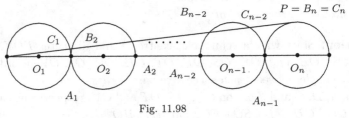

Fig. 11.98

Solution. Draw perpendiculars from O_k to the tangent line, meeting at M_k ($k = 1, \ldots, n$). These will meet the tangent line at the mid-points of the line segments B_kC_k.

Also, $\triangle B_kO_kM_K$ is a right triangle, with $O_kB_k = r$.

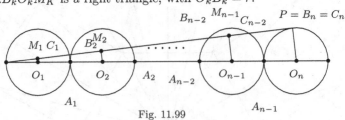

Fig. 11.99

Note that $A_0O_1 : A_0O_2 : \cdots A_0O_{n-1} : A_0O_n = 1 : 3 : \cdots 2n-3 : 2n-1$. By similar triangles, we obtain that

$$M_1O_1 : M_2O_2 : \cdots M_{n-1}O_{n-1} : M_nO_n = 1 : 3 : \cdots 2n - 3 : 2n - 1.$$

Remember that $M_nO_n = r$. This shows that $M_kO_k = \left(\frac{2k-1}{2n-1}\right)r$.

Then, $(B_kM_k)^2 = (B_kO_k)^2 - (M_kO_k)^2 = r^2\left(1 - \left(\frac{2k-1}{2n-1}\right)^2\right)$,

so that $B_kC_k = 2r\sqrt{1 - \left(\frac{2k-1}{2n-1}\right)^2}$.

Comment: the classic example is when $n = 3$, $k = 2$ and $r = 5$, giving a chord length of 8.

Problem 11.150. *The line with slope $\lambda > 0$ acts like a mirror to a ray of light coming along a line parallel to the x-axis from $+\infty$. Determine the slope of the reflected ray.*

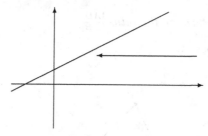

Fig. 11.100

Solution. We know that the angle between the incident ray and the mirror and the angle between the reflected ray and the mirror are equal. If we construct a right triangle for the incident ray, we need to find a similar triangle for the reflected ray.

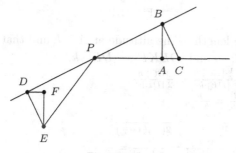

Fig. 11.101

In fact, we construct two (similar) triangles for the incident ray: let P be the point where the incident ray meets the mirror and B a point on the mirror to the right of P. Draw $BA \perp PA$ (A on the incident ray) and $PB \perp PB$ (C on the incident ray). Let $AC = 1$. Then $BA = \lambda$, $PA = \lambda^2$, and $PB = \lambda\sqrt{1 + \lambda^2}$.

Let D be the symmetric point on the mirror to B through P, so that $DP = \lambda\sqrt{1 + \lambda^2}$. Let $DF \parallel PC$ and $DF = AC$. Let $DE \perp DP$ and $FE \perp DF$. Then $FE = \lambda$ and $DE = \sqrt{1 + \lambda^2}$.

Thus, $\triangle PDE$ is congruent to $\triangle PBC$, and $\angle BPC = \angle DPE$. If we let P be the origin, it is easy to find that the slope of the reflected ray is $\frac{2}{1-\lambda}$ when $\lambda \neq 1$, in which case, the reflected ray is parallel to the y–axis.

Problem 11.151. *Given triangle ABC and DE $\parallel$ BC, with D $\in$ AB and E $\in$ AC. Drop perpendiculars from D and E to BC, meeting BC at F and K, respectively.*

If $\frac{[ABC]}{[DEKF]} = \frac{32}{7}$, determine the ratio $\frac{|AD|}{|DB|}$.

Solution. Let $\frac{|AD|}{|DB|} = T$.

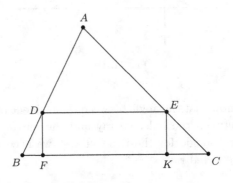

Fig. 11.102

Suppose that the length of the altitude at A is h and that $FD = k$. Then $[ABC] = \frac{1}{2} \cdot |BC| \cdot h$ and $[DEKF] = |DE| \cdot k$.

Then $\frac{[ABC]}{[DEKF]} = \frac{\frac{1}{2} \cdot |BC| \cdot h}{|DE| \cdot k} = \frac{|BC| \cdot h}{2 \cdot |DE| \cdot k}$.

Now, $\frac{|BC|}{|DE|} = \frac{|AB|}{|AD|} = \frac{h}{h-k}$,

giving $\frac{[ABC]}{[DEKF]} = \frac{h^2}{2k(h-k)} = \frac{1}{2\left(\frac{k}{h}\right)\left(1 - \frac{k}{h}\right)}$.

Let $x = \frac{k}{h}$. Thus, $\frac{[ABC]}{[DEKF]} = \frac{1}{x(1-x)} = \frac{32}{7}$.

Solving for x gives $x = \frac{1}{8}$ or $x = \frac{7}{8}$.

Now, $x = \frac{k}{h} = \frac{|BC|}{|AB|} = \frac{|BD|}{|AD|+|DB|} = \frac{1}{T+1}$.

Solving this gives answers of $T = 7$ or $\frac{1}{7}$.

Problem 11.152. *Determine the maximum area of an isosceles triangle inscribed in a circle of radius r*

Solution. The area of the triangle is $x(\sqrt{r^2 - x^2} + r)$.

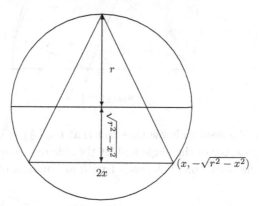

Fig. 11.103

Differentiate this to obtain $\sqrt{r^2 - x^2} - \frac{x^2}{\sqrt{r^2 - x^2}} + r$.

Set this equal to zero to find candidates for a maximum or minimum:
$\sqrt{r^2 - x^2} - \frac{x^2}{\sqrt{r^2 - x^2}} + r = 0$.

It helps to multiply by $\sqrt{r^2 - x^2}$. This leads to

$$\left(\sqrt{r^2 - x^2}\right)\left(\sqrt{r^2 - x^2} - \frac{x^2}{\sqrt{r^2 - x^2}} + r\right) = r\sqrt{r^2 - x^2} - 2x^2 + r^2 = 0 .$$

Now, re-arrange to get $r\sqrt{r^2 - x^2} = 2x^2 - r^2$.

Square both sides, to get $\left(r\sqrt{r^2 - x^2}\right)^2 = \left(2x^2 - r^2\right)^2$.

Moving all terms to the left and expanding gives

$$3r^2x^2 - 4x^4 = x^2\left(3r^2 - 2x^2\right) = 0 .$$

Solving gives $x = 0$, $x = \frac{r\sqrt{3}}{2}$, and $x = -\frac{r\sqrt{3}}{2}$.

The last of these is an impossible value.

Investigation shows that $x = 0$ gives a minimum and $x = \frac{r\sqrt{3}}{2}$ gives the required maximum.

Substituting into the area formula gives the maximum value of $\frac{3r^2\sqrt{3}}{4}$.

Problem 11.153. *Square ABCD is inscribed in one eighth of a circle of radius 1, such that there is one vertex on each radius and two vertices on the arc.*

Determine the area of the square. Exact answer required in the form $\frac{a + b\sqrt{c}}{d}$, where a, b, c and d are integers.

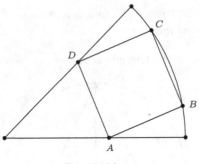

Fig. 11.104

Solution. We need to know that fact that $\tan\left(\frac{\pi}{8}\right) = \sqrt{2} - 1$.
Draw the bisector of the angle and let the side of the square be $2a$. Let the
centre of the sector be O. Various length are marked on the diagram.

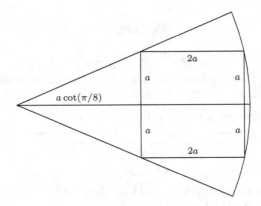

Fig. 11.105

We have $(2a)^2 + (2a + a\cot(\pi/8))^2 = 1$, which yields $a = \sqrt{5/51 - 2\sqrt{2}/51}$.
Thus, the area of the square is $4a^2 = \frac{20 - 8\sqrt{2}}{51}$.

Problem 11.154. *A right angled triangle ABC has the following property:
draw a square externally on each side; the vertices of the squares not on the
right triangle are concyclic.*
Characterize such triangles.

Solution. We draw a diagram. If the six points are concyclic, then the
outer sides of the squares are chords of the same circle; their perpendicular
bisectors will be concurrent.

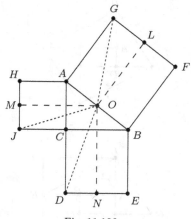

Fig. 11.106

It is easy to see that these three perpendicular bisectors are concurrent for any right triangle; they are the perpendicular bisectors of the sides, which concur at the circumcentre O.

But we must determine that, WLOG, $OJ = OD = OG$.

Let the sides of $\triangle ABC$ be a, b and c as usual.

Then, $OJ^2 = OM^2 + MJ^2 = \left(b + \frac{a}{2}\right)^2 + \left(\frac{b}{2}\right)^2$

and $OD^2 = ON^2 + ND^2 = \left(a + \frac{b}{2}\right)^2 + \left(\frac{a}{2}\right)^2$.

For these to be equal, we get $a^2 = b^2$, giving $a = b$ (since both are positive).

Thus, $OJ^2 = OD^2 = \frac{5a^2}{2}$.

Now, $OG^2 = OL^2 + LG^2 = c^2 + \left(\frac{c}{2}\right)^2 = \frac{5c^2}{4} = \frac{5(a^2+b^2)}{5} = \frac{5a^2}{5}$, since $a = b$.

Thus, the right triangles with this property are the isosceles right triangles.

Problem 11.155. *The sides of rectangle $ABCD$ are extended as follows:*

AB is extended to E so that $AB = BE$.

BC is extended to F so that $FC = 2CB$.

CD is extended to G so that $CD = DG$.

DA is extended to H so that $HA = 2AB$.

Determine the value of the fraction $\dfrac{area\ of\ EFGH}{area\ of\ ABCD}$.

Solution. The simplest solution involves copying triangle CFG onto EH and triangle DHG onto EF.

This means that the area of quadrilateral $EFGH$ is the same as the area of polygon $AHC'B'FCD$.

Then, divide rectangles $AHC'E$ and $BEB'F$ into copies of rectangle $ABCD$.

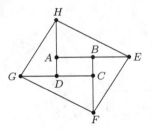

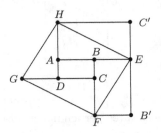

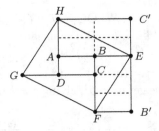

We now see that the area of polygon $AHC'B'FCD$ consists of 8 copies of rectangle $ABCD$.

Hence, the value of the fraction $\dfrac{\text{area of } EFGH}{\text{area of } ABCD}$ is 8.

Problem 11.156. *In triangle ABC, angle BAC is a right angle and $AC > AB$. The mid-points of BC, AC, and AB are L, M, and N, respectively.*

The circle, centre L, radius $\frac{AB}{2}$ intersects BL at P and CL at Q.

If $PN \parallel QM$, determine the size of angle ACB.

Solution. Let $AB = 2x$. $AC = 2y$, and $BC = 2z$.

Then $PL = LQ = LM = x$, $AM = MC = NL = y$, and $BP = CQ = z - x$.

Since $NL \parallel AC$, we have $\angle PLN = \angle QCM$.

Since $PN \parallel QC$, we have $\angle PNL = \angle QMC$.

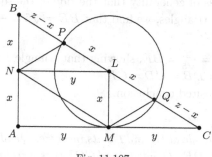

Fig. 11.107

Thus, triangles PNL and QMC are similar. Since $BN = LM$, the triangles are, in fact, congruent.

Therefore, $PL = QC$; that is, $x = z - x$, or $z = 2x$, showing that $BC = 2AB$.

From this, we deduce that $\angle BCA = 30°$.

Problem 11.157. *In square $ABCD$ is the mid-point of AC, F is the point on B such that $DF \perp EB$, and G is the point on EF such that $AG \perp EB$. Show that $CF = DG$*

Solution. Draw a diagram and join ED.

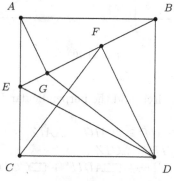

Fig. 11.108

Note that quadrilateral $CEFD$ is cyclic. Hence, $\angle EFC = \angle EDC$.
Note that $\angle EDC = \angle ABE = \angle FDB = \theta$, say, since
$$\triangle BFD \sim \triangle EBA \cong \triangle EDE .$$
Hence, $\angle CFD = \angle CDF = 90° - \theta$. Thus, $\triangle CDF$ is isosceles with $CF = CD$.

Assume without loss of generality that the side of the square is 2.
Then, using similar triangles, we find that $BE = \sqrt{5} = \frac{5}{\sqrt{5}}$, $BF = \frac{2}{\sqrt{5}}$, and $EG = \frac{1}{\sqrt{5}}$.

Then, we have $GF = \frac{2}{\sqrt{5}} = BF$, showing that triangle BEG is an isosceles triangle with $DG = DB = CD = CF$.

We now have the desired conclusion.

Problem 11.158.

$\triangle ABC$ has a right angle at B, and I is its incentre. Quadrilateral $ADEB$ is a rectangle with $I \in DE$. Quadrilateral $CGFB$ is a rectangle with $I \in FG$. Let $H = DE \cap AC$, and $K = FG \cap AC$.

Prove that

$$\triangle HIK = \triangle ADH + \triangle CGK .$$

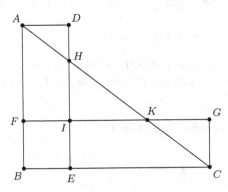

Fig. 11.109

Solution. Draw the incircle and the perpendicular IZ from I to AC. Join AI and CI.

It is easy to see that $\triangle AID \cong \triangle AIG \cong \triangle AIZ$
and that $\triangle CIG \cong \triangle CIE \cong \triangle CIZ$.

Hence, $\triangle ABC = \square BEIF + \square ADIF + \square CGIE$.

The result follows.

Problem 11.159. *Triangle ABC with $a = BC \geq CA = b$ has a right angle at C. Squares $ABDE$, $BCFG$ and $CAHI$ are drawn externally to triangle ABC as shown.*

Let $FI \cap EH = P$, $IF \cap DG = Q$ and $GD \cap HE = R$.

Determine the ratio $\frac{b}{a}$ such that triangle PQR is a right triangle.

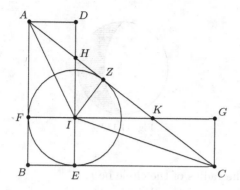

Fig. 11.110

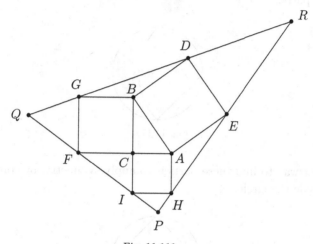

Fig. 11.111

Solution. Let $A = (b, 0)$, $B = (0, a)$ and $C = (0, 0)$.

It is then straightforward to find the coordinates of D, E, F, G, H and L, and, further, the slopes of FI and HE.

This results in the ratio $\frac{b}{a} = \frac{1}{\sqrt{2}}$.

Problem 11.160. *On a certain day, the moon is seen with the shadow passing through diametrically opposite points. A photograph is shown.*

Calculate the exact proportion of the moon that is bright as seen on the photograph. See Fig. 11.112.

HINT: Remember that the shadow of a sphere appears as a circle.

Fig. 11.112

Solution. Let the radius of the circle be r.
The secret here is to note that bright area consists of a quarter circle plus two other parts.

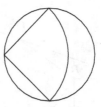

Fig. 11.113

The easiest way to find these parts is to apply symmetry, by completing a square inside the circle.

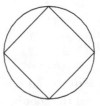

Fig. 11.114

The sum of the four equal parts is the area of the circle less the area of the square; that is $\pi r^2 - (r\sqrt{2})^2 = r^2(\pi - 2)$. Hence, the sum of the two parts is one-half of this; that is $\frac{r^2(\pi-2)}{2}$.

The area of the quarter circle is $\frac{1}{4}\pi(r\sqrt{2})^2 = \frac{\pi r^2}{2}$.

Hence, the required proportion is $\frac{\frac{r^2(\pi-2)}{2} + \frac{\pi r^2}{2}}{\pi r^2} = \boxed{\frac{\pi-1}{\pi}} = \boxed{1 - \frac{1}{\pi}}$.

Problem 11.161. *Two squares XZBA and YZCD are drawn outside △XYZ.*

Prove that the point of intersection of XD and YA lies on the altitude of △XYZ which passes through the vertex Z.

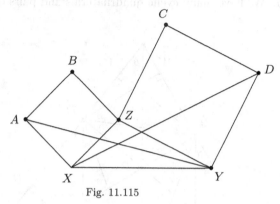

Fig. 11.115

Solution. Draw △AXY ≅ △XZP. This implies that $XP ⊥ AY$. By symmetry, △YZP ≅ △DYX. This implies that $YP ⊥ XD$.

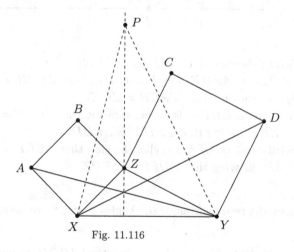

Fig. 11.116

Hence, XD and YA are altitudes of △XYP, showing that PZ is also an altitude of △XYP. Thus, $PZ ⊥ XY$, and therefore, passes through the orthocentre of △XYZ. This proves the result.

Problem 11.162. (This is Problem 9.4 on page 128.) *Let H be the orthocentre of $\triangle ABC$. Prove that this diagram is legitimate.*

Solution. We have many cyclic quadrilaterals and pairs of parallel lines.

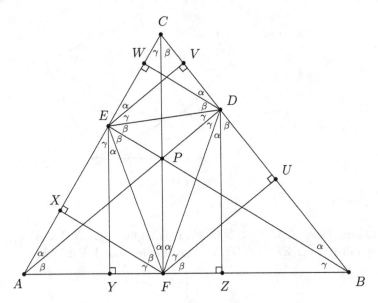

(1) $CF \| EY$, showing that $\angle YEZ = \angle ECH$;
(2) Quadrilateral $AFHE$ is cyclic, showing that $\angle EFH = \angle EAH$;
(3) $AD \| EV$, showing that $\angle AEH = \angle CEV$;
(4) Quadrilateral $EDVW$ is cyclic, showing that $\angle CEV = \angle CDW$;
(5) $WD \| EB$, showing that $\angle CDW = \angle EBD$;
(6) Quadrilateral $BDHF$ is cyclic, showing that $\angle EBD = \angle DFC$;
(7) $CF \| ZB$, showing that $\angle DFC = \angle FDZ$.

This shows the result for angle α. Angles β and γ are done similarly.

Problem 11.163. *Given cyclic quadrilateral $ABCD$, suppose that the lines AB and CD meet at P, and that the line AD and BC meet at Q. The internal bisector of $\angle AQB$ meets DC and AB at G and E, respectively. The internal bisector of $\angle APD$ meets BC and AD at F and H, respectively. Prove that $EFGH$ is a rhombus.*

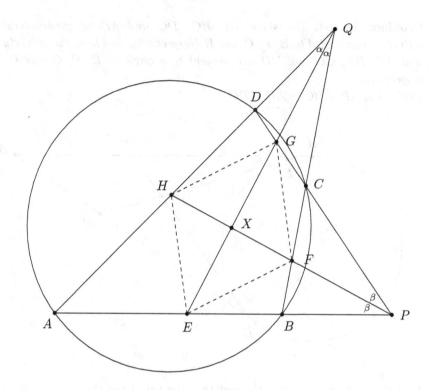

Solution. We make use of the Sine Rule.

We have $\dfrac{AE}{\sin(\alpha)} = \dfrac{EQ}{\sin(\widehat{A})}$ and $\dfrac{ED}{\sin(\alpha)} = \dfrac{EQ}{\sin(\widehat{B})}$. Therefore, $\dfrac{AE}{ED} = \dfrac{\sin(\widehat{B})}{\sin(\widehat{A})}$.

Similarly, $\dfrac{AH}{HD} = \dfrac{\sin(\widehat{D})}{\sin(\widehat{A})}$. But, $\widehat{B} + \widehat{D} = 180°$, so that $\sin(\widehat{B}) = \sin(\widehat{D})$.

Hence, $\dfrac{AE}{ED} = \dfrac{AH}{HD}$, implying that $HE\|DB$. Similarly, $HG\|EF$, showing that quadrilateral $EFGH$ is a parallelogram.

To complete the proof, it is sufficient to show that $\angle QXP = 90°$.

First, note that $\angle PBQ = \widehat{D}$, so that $\angle PFQ = \widehat{D} + \beta$, which gives that $\angle QFX = 180° - \widehat{D} - \beta$.

Therefore, $\angle QXF = 180° - (180° - \widehat{D} - \beta) - \alpha = \widehat{D} + \beta - \alpha$.

From $\triangle CDQ$, we have $\widehat{B} + \widehat{A} + 2\alpha = 180°$, and, from $\triangle BCP$, that $\widehat{A} + \widehat{D} + 2\beta = 180°$.

Subtracting these equations gives $\beta - \alpha = \dfrac{\widehat{B} - \widehat{D}}{2}$.

Thus, $\angle QXF = \widehat{D} + \left(\dfrac{\widehat{B} - \widehat{D}}{2}\right) = \dfrac{\widehat{D} + \widehat{B}}{2} = 90°$.

Problem 11.164. *The sides AB, BC, DC and AD of quadrilateral ABCD are extended to E, F, G and H, respectively, and have the property that AE, BF, CG and DH are tangent to a circle at E, F, G and H, respectively.*
Prove that $AB + BC = AD + DC$.

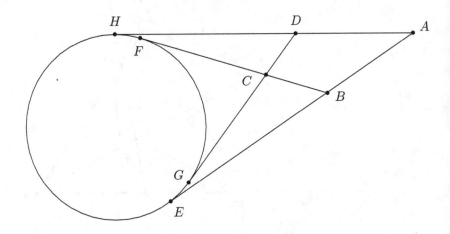

Solution. We make use of the fact that the lengths of the tangents from an exterior point to a circle are equal.

$$AB + BC = (AE - BE) + (BF - CF)$$
$$= AE - CF$$

since $BE = BF$,

$$= AH - CG$$

since $AH = AE$ and $CF = CG$,

$$= (AD + DH) - (DG - DC)$$
$$= AD + DC$$

since $DH = DG$.

Problem 11.165. *The internal bisectors of angles $\angle ACB$ and $\angle CAB$ meet the opposite sides AB and CB at E and D, respectively.*
Let F be an point on the line segment ED. Drop perpendiculars FG, FH and FN to sides AB, BC and CA, respectively.
Prove that $FG + FH = FN$.

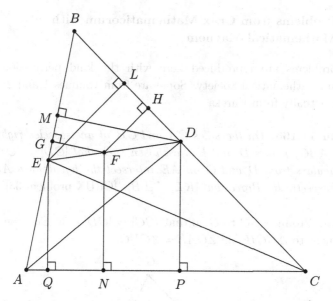

Solution. Drop perpendiculars DM and DP to sides AB and AC, respectively. Drop perpendicualrs EL and EQ to sides BC and AC, respectively. Suppose that DE and CA intersect at X. Let $x = EF$, $y = FD$ and $EX = l$. Then from similar triangles, we have

$$\frac{AE}{XQ} = \frac{XF}{XN} = \frac{XD}{XP}.$$

Then we have

$$\frac{EL}{FH} = \frac{ED}{FD} \qquad \frac{DM}{FG} = \frac{ED}{EF},$$

giving

$$FH = EL \cdot \frac{FD}{ED} = EQ \cdot \frac{FD}{ED},$$

$$FG = DM \cdot \frac{EF}{ED} = DP \cdot \frac{EF}{ED}.$$

From these, it follows that

$$\frac{FH + FG}{FN} = \left(\frac{EQ}{FN} \cdot \frac{FD}{ED} + \frac{DP}{FN} \cdot \frac{EF}{ED} \right)$$

$$= \left(\frac{l}{l+x} \cdot \frac{y}{x+y} + \frac{l+x+y}{l+x} \cdot \frac{x}{x+y} \right)$$

$$= 1 \qquad \text{by elementary algebra.}$$

11.1   Problems from Crux Mathematicorum with Mathematical Mayhem

These problems are reproduced here with the kind permission of the Canadian Mathematical Society. Some are from volumes 1 and 2; so that they are actually from Eureka.

Problem 11.166. *On the sides CA and CB of an isosceles right-angled triangle ABC, points D and E are chosen such that $|CD| = |CE|$. The perpendiculars from D and C on AE intersect the hypotenuse AB in K and L, respectively. Prove that $|KL| = |LB|$.* [CRUX problem 33]

Solution. Produce DC to G so that $|CG| = |CD| = |CE|$; then $\triangle ACE$ is congruent to $\triangle GCB$ and $\angle CAE = \angle CBG$.

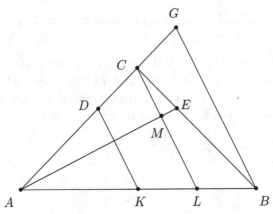

Fig. 11.117

But if CL meets AE in M, then $\triangle ACE$ is similar to $\triangle CME$ and thus, $\angle MCE = \angle CAE = \angle CBG$, from which $GB\|CL\|DK$. Since C bisects DG, it follows that L bisects KB.

Problem 11.167. *From the centres of each of two non-intersecting circles tangents are drawn to the other circle, as shown in the diagram below. Prove that the chords PQ and RS are equal in length. (It is reputed that this problem originated with Newton, but I have not been able to find an exact reference.)* [CRUX problem 63]

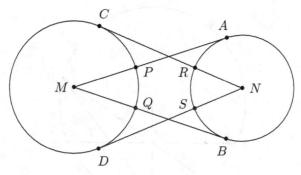

Fig. 11.118

Solution. The line of centres MN intersects the chords PQ and RS at their mid-points F and G. It is clear that

$$\triangle MFP \sim \triangle MAN \quad \text{and} \quad \triangle NGR \sim \triangle NCM .$$

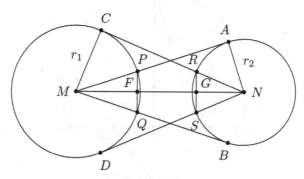

Fig. 11.119

Hence, if r_1 and r_2 denote the radii, as shown in Fig. 11.119, we have

$$\frac{1}{2}PQ = FP = \frac{r_1 r_2}{MN} \quad \text{and} \quad \frac{1}{2}RS = GR = \frac{r_1 r_2}{MN} ,$$

and thus, $PQ = RS$.

Problem 11.168. *Suppose that M is the mid-point of chord AB of the circle with centre C as shown in Fig. 11.120. Prove that $RS > MN$.* [CRUX problem 75]

Solution. Join ND and NS, where D is the extremity of the diameter through S. Two pairs of the equal angles are denoted by α and β, and $x = \frac{\pi}{2} - \beta$, $y = \frac{\pi}{2} - \alpha$.

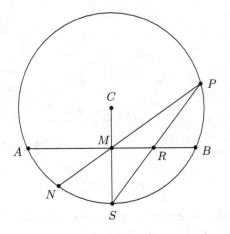

Fig. 11.120

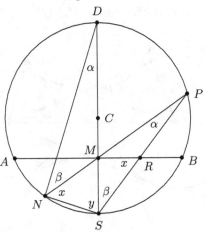

Fig. 11.121

From triangle's MNS and MSR, we have

$$\frac{MN}{MS} = \frac{\sin y}{\sin x} = \frac{\cos \alpha}{\sin x} \quad \text{and} \quad \frac{MS}{RS} = \sin x \ ,$$

so that

$$\frac{MN}{RS} = \frac{MN}{MS} \cdot \frac{MS}{RS} = \cos \alpha \leq 1 \ ,$$

and thus, $RS \geq MN$, equality being attained only when P is at D.

Problem 11.169. *Show that, for any triangle ABC,*

$$|OA|^2 \sin A + |OB|^2 \sin B + |OC|^2 \sin C = 2K \ ,$$

where O is the centre of the inscribed circle and K is the area of △ABC.
[CRUX problem 126]

Solution. By looking at Fig. 11.122, we see that

$$2[APOR] = 4[APO] = 2 \cdot |OP| \cdot |AP|$$

$$= 2 \cdot |OP| \sin \frac{A}{2} \cdot |OA| \cos \frac{A}{2} = |OA|^2 \sin A \ .$$

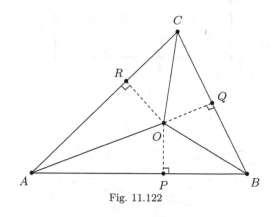

Fig. 11.122

Also, by replacing A successively by B and C, and adding, we obtain

$$2K = |OA|^2 \sin A + |OB|^2 \sin B + |OC|^2 \sin C \ .$$

Problem 11.170. *In square ABCD, $\overline{AC}$ and $\overline{BD}$ meet at E. Point F is in $\overline{CD}$ and $\angle CAF = \angle FAD$. If $\overline{AF}$ meets $\overline{ED}$ at G and if $EG = 24$, find CF.* [CRUX problem 147]

Solution. Draw $EK \| DC$ as shown in Fig. 11.123.
In right triangle AEG, we have $\angle AGE = 67\frac{1}{2}°$; and $\angle EKG = \angle KAE + \angle AEK = 67\frac{1}{2}°$. Hence, triangle KEG is isosceles and

$$FC = 2KE = 2EG = 48 \ .$$

Addendum.

Show, more generally, that if $ABCD$ is a rectangle with $AB : BC = m$ and if $EG = k$, then

$$CF = 2k \left(1 + \frac{m^2 - 1}{\sqrt{m^2 + 1}} \right) \ .$$

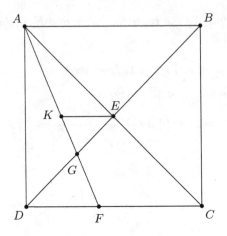

Fig. 11.123

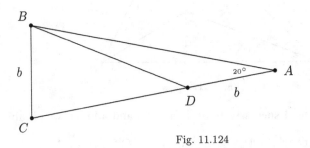

Fig. 11.124

Problem 11.171. *Consider the isosceles triangle ABC in the figure, which has a vertical angle of* 20°. [CRUX problem 175]

On AC, one of the equal sides, a point D is marked off so that |AD| = |BC| = b. *Find the measure of* ∠ABD.

Solution. I. If △ABE is equilateral (see Fig. 11.125),
then △EAD = △ABC (SAS) and △EBD is isosceles.

Since ∠BED = 60° − 20° = 40°, it follows that ∠EBD = 70° and ∠ABD = 10°.

II. An unexpected by–product of this fascinating triangle is shown in Fig. 11.126.

By symmetry, ADEB is an isosceles trapezoid; hence, ∠CDE = 20° and DE = EC = b.

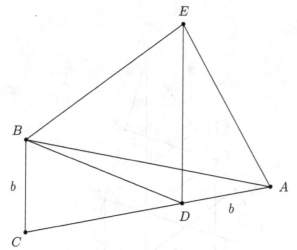

Fig. 11.125

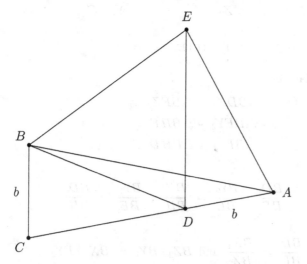

Fig. 11.126

Problem 11.172. *Let D, E, F denote the feet of the altitudes of $\triangle ABC$, and let (X_1, X_2), (Y_1, Y_2), (Z_1, Z_2) denote the feet of perpendiculars from D, E, F, respectively, upon the other two sides of the triangle. Prove that the six points X_1, X_2, Y_1, Y_2, Z_1, Z_2 lie on a circle.* [CRUX problem 192]

Solution. Let H be the orthocentre of $\triangle ABC$, and let Z_2 and Y_1 lie on BC, X_2 and Z_1 on CA, and Y_2 and X_1 on AB, as shown in the Fig. 11.127.

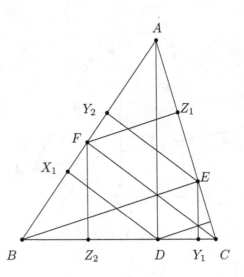

Fig. 11.127

From the similarities,

$$\triangle BDX_1 \sim \triangle BFZ_2 \ ,$$
$$\triangle BEY_2 \sim \triangle BHF \ ,$$
$$\triangle BEY_1 \sim \triangle BHD \ ,$$

we get

$$\frac{BX_1}{BZ_2} = \frac{BD}{BF} \ , \ \frac{BY_2}{BE} = \frac{BF}{BH} \ , \ \frac{BY_1}{BE} = \frac{BD}{BH} \ ,$$

so that

$$\frac{BY_1}{BY_2} = \frac{BD}{BF} = \frac{BX_1}{BZ_2} \text{ and } BZ_2 \cdot BY_1 = BX_1 \cdot BY_2 \ .$$

Hence, Y_2, X_1, Z_2, $Y - 1$ lie on a circle γ_1; and similarly Z_2, Y_1, X_2, Z_1 lie on a circle γ_2; and X_2, Z_1, Y_2, X_1 lie on a circle γ_3.

If γ_1 and γ_2, for example, coincide, our theorem is proved; and if they are distinct their radical axis is the line BC. Thus, if our theorem is not true, we have three distinct circle γ_1, γ_2, and γ_3, whose radical axes, taking the circles in pairs, form the sides of $\triangle ABC$, which contradicts the theorem that such radical axes are either concurrent or parallel (see D. Pedoe, *A Course of Geometry for Colleges and Universities*, Cambridge University Press, London, 1970, p. 110, Theorem 28.1, for example).

Comment.

The circle in this problem is known in the literature as *The Taylor Circle*, after H.M. Taylor (1842–1927) who discussed it in "On a six–point circle connected with a triangle", *The Messenger of Mathematics*, new series, Vol. XI, 1881–1882, pp. 177–179. In spite of its name, this circle did not originate with Taylor. It was discovered earlier, like so many other good things in life, by the French. It was mentioned by Eugène Catalan, *Théorèmes et problèmes de Géométrie élémentaire*, sixième édition, Paris, 1879, livre III, ch. LXV, p. 132, and was apparently first proposed by the French mathematician known as Eutaris (pen name (anagram) of Restiau, of Collège Chaptal, Paris) in *Journal de Mathématiques élémentaires* de M. Vuibert, Vol. 2, 1877, pp. 30, 43, No. 60.

Problem 11.173. *If a quadrilateral is circumscribed about a circle, prove that its diagonals and the two chords joining the points of contact of opposite sides are all concurrent.* [CRUX problem 199]

Solution. This theorem is true not only for a circle but for any conic.

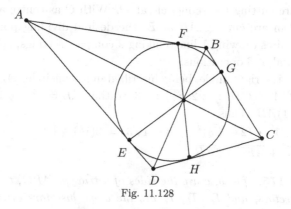

Fig. 11.128

Brianchon's Theorem states that the lines joining the opposite vertices of a hexagon circumscribed about a conic are concurrent. The quadrilateral in Fig. 11.128 can be considered as a degenerate circumscribed hexagon in two ways: either $AFBCHD$, whereupon FH passes through P, the intersection of AC and BD; or $ABGCDE$, whereupon AC, BD, and GE are concurrent at P. Hence, the diagonals and the joins of the opposite points of tangency of the quadrilateral are concurrent.

Problem 11.174. *Devise a Euclidean construction to divide a given line segment into two parts such that the sum of the squares on the whole segment and on one of its parts is equal to twice the square on the other part.* [CRUX problem 158]

Solution. *Analysis.* If the length of the given segment is y and that of the first part is x, then $x^2 + y^2 = 2(y-x)^2$ or $x^2 - 4xy + y^2 = 0$. Since $x < y$, the applicable solution of this equation is $x = y(2 - \sqrt{3})$.

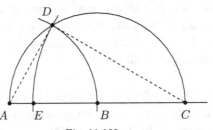

Fig. 11.129

Construction. Extend the given segment AB. With radius BA and B as centre describe semicircle AC. With the same radius and A as centre describe an arc cutting the semicircle at D. With C as centre and radius CD describe an arc cutting AB at E, the desired dividing point. (This procedure requires drawing 1 line, making 2 compass settings, and striking 3 arcs — a total of 6 operations.)

Proof. $\angle ADC$ is a right angle, being inscribed in a semicircle. $AD = AB = \frac{AC}{2}$. Hence, $CA : CD : : 2 : 1 : \sqrt{3}$, so that $AE = (2 - \sqrt{3})AB$, and $EB = (\sqrt{3} - 1)AB$. Thus,

$$1^2 + (2 - \sqrt{3})^2 = 8 - 4\sqrt{3} = 2(\sqrt{3} - 1)^2 \ ,$$

and thus, $\overline{AB}^2 + \overline{AE}^2 = 2\overline{EB}^2$.

Problem 11.175. *If a, b, c are the sides of a triangle ABC, t_a, t_b, t_c are the angle bisectors, and T_a, T_b, T_c are the angle bisectors extended until they are chords of the circle circumscribing the triangle ABC, prove that* [CRUX problem 168]

$$abc = \sqrt{T_a T_b T_c t_a t_b t_c} \ .$$

Solution. If the bisector of $\angle A$ meets the opposite side at D and the circumcircle at E (see figure), then $\triangle$'s ABC and AEC are similar, and thus, $bc = T_a t_a$. Similarly $ca = T_b t_b$, $ab = T_c t_c$, and the stated result follows immediately.

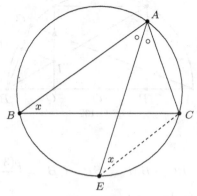

Fig. 11.130

Problem 11.176. *A square PQRS is inscribed in a semicircle (O) with PQ falling along diameter AB (see figure). A right triangle ABC, equivalent to the square, is inscribed in the same semicircle with C lying on the arc RB. Show that the incentre I of triangle ABC lies at the intersection of SB and RQ, and that*

$$\frac{RI}{IQ} = \frac{SI}{IB} = \frac{1+\sqrt{5}}{2} \ ,$$

the golden ratio. [CRUX problem 368]

Solution. Let $BS \cap QR = I$. Since $\triangle ROS = \triangle BOC$, we have arc $RS =$ arc BC and

$$\text{arc } CS \ = \ \text{arc } BR \ = \ \text{arc } SA \ .$$

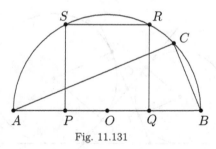

Fig. 11.131

Thus, BS bisects $\angle ABC$ since S bisects arc CA. Drop perpendicular IM on BC. If $OR = \rho$, then

$$SP = \frac{2\rho}{\sqrt{5}}, \quad QB = \rho\left(1 - \frac{1}{\sqrt{5}}\right), \quad PB = \rho\left(1 + \frac{1}{\sqrt{5}}\right) \ .$$

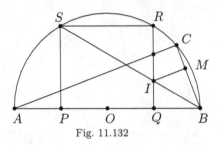

Fig. 11.132

We have

$$MC = BC - BM = SP - QB = \frac{\rho(3 - \sqrt{5})}{\sqrt{5}}$$

and, from similar triangles,

$$IM = IQ = QB \cdot \frac{SP}{PB} = \frac{\rho(3 - \sqrt{5})}{\sqrt{5}} \; .$$

Thus, $IM = MC$ and IC bisects $\angle ACB$, which makes I the incentre of $\triangle ABC$. Finally, from similar triangles,

$$\frac{RI}{IQ} = \frac{SI}{IB} = \frac{SR}{QB} = \frac{SP}{QB} = \frac{1 + \sqrt{5}}{2} \; .$$

Problem 11.177. *In $\triangle ABC$, the centres of the escribed circles opposite A and B are at X and Y, respectively. Prove that X, C, and Y are collinear.* [MAYHEM problem HS4]

Solution. Let D, E be the points where BC and AC are tangent to circle Y and circle X, respectively, and let G and F be the points where BC and AC are tangent to circle X and circle Y, respectively, as shown.

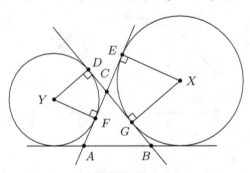

Fig. 11.133

In triangles XGC and XEC, we have $XE = XG$, $\angle XEC = \angle XGC$, and XC is common. Hence, $\triangle XGC \cong \triangle XEC$.

Thus, $\angle XCE = \angle XCG$. Similarly, $\angle YCD = \angle YCF$.
Finally, since $\angle DCF = \angle ECG$, we have $\angle XCE = \angle XCG = \angle YCD = \angle YCF$, and since $\angle XCE + \angle XCG + \angle GCA = 180°$, $\angle XCG + \angle GCA + \angle YCF = 180°$ (since $\angle XCE = \angle YCF$), and XCY is a straight line. We conclude that X, C, and Y are collinear, as required.

Problem 11.178. *Prove that, for any parallelogram $ABCD$, where A, B, C, and D are consecutively labelled vertices,*

$$AB^2 + BC^2 + CD^2 + DA^2 = AC^2 = BD^2.$$

[MAYHEM problem HS6]

Solution. Let a, b, and h be as indicated. Then:

$$AB^2 = CD^2 = a^2,$$
$$AD^2 = BC^2 = h^2 + b^2,$$

hence, $AB^2 + BC^2 + CD^2 + DA^2 = 2a^2 + 2b^2 + 2h^2$.

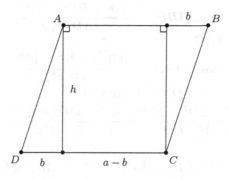

Fig. 11.134

Now, $BD^2 = (a+b)^2 + h^2$ and $AC^2 = (a-b)^2 + h^2$. Thus,

$$AC^2 + BD^2 = (a+b)^2 + (a-b)^2 + 2h^2 = 2a^2 + 2b^2 + 2h^2.$$

Therefore, $AB^2 + BC^2 + CD^2 + DA^2 = AC^2 + BD^2$.
ALTERNATIVELY:
Since $ABCD$ is a parallelogram, $\angle ABC + \angle BCD = 180°$, so that $\cos \angle BCD = -\cos \angle ABC$.

Now, by the Cosine Law, we have

$$AC^2 = AB^2 + BC^2 - \frac{1}{2}AB \cdot BC \cos \angle ABC,$$

$$BD^2 = BC^2 + CD^2 - \frac{1}{2}BC \cdot CD \cos \angle BCD$$

$$= AD^2 + CD^2 + \frac{1}{2}AB \cdot BC \cos \angle ABC.$$

Adding gives

$$AC^2 + BD^2 = AB^2 + BC^2 + CD^2 + DA^2.$$

Problem 11.179. *In* $\triangle ABC$, *the bisector of* $\angle A$ *intersects* BC *at* D. *Prove that*

$$AD^2 = AB \cdot AC - BD \cdot CD.$$

[MAYHEM problem HS17]

Solution. By the Law of Cosines in $\triangle ABD$,

$$AB^2 = AD^2 + BD^2 - 2 \cdot AD \cdot BD \cdot \cos \angle ADB$$

so that

$$\cos \angle ADB = \frac{AD^2 + BD^2 - AB^2}{2 \cdot AD \cdot BD}.$$

Similarly, from $\triangle ADC$,

$$\cos \angle ADC = \frac{AD^2 + CD^2 - AC^2}{2 \cdot AD \cdot CD}.$$

Since $\cos \angle ADB + \cos \angle ADC = 0$, we have

$$\frac{AD^2 + BD^2 - AB^2}{2 \cdot AD \cdot BD} + \frac{AD^2 + CD^2 - AC^2}{2 \cdot AD \cdot CD} = 0.$$

Thus,

$$AD^2(BD + CD) + BD \cdot CD(BD + CD) = AB^2 \cdot CD + AC^2 \cdot BD$$

$$\text{or} \quad BC(AD^2 + BC \cdot CD) = AB^2 \cdot CD + AC^2 \cdot BD.$$

(This is Stewart's Theorem, which is valid for any cevian AD.)
Now, since AD is an angle bisector, $AB : AC = BD : CD$, giving $AB \cdot CD = AC \cdot BD$.
Using this fact, we have

$$BC(AD^2 + DB \cdot CD) = AB(AC \cdot BD) + AC(AB \cdot CD)$$

$$= AB \cdot AC \cdot BC.$$

Thus, we conclude that $AD^2 = AB \cdot AC - BD \cdot DC$, as required.

Problem 11.180. *In △ABC, AB = AC. If D is on BC extended such that BC = CD and E is on AB extended, so that AB = BE, show that AD = CE.* [MAYHEM problem HS20]

Solution. In triangles ACD and EBC, we have

$$AC = EB, \qquad \angle ACD = \angle EBD, \qquad CD = BC.$$

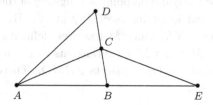

Thus, by side-angle-side congruence, $\triangle ACD = \triangle EBC$, from which $EC = AD$, as required.

Problem 11.181. *Let D and E be points on sides AC and AB respectively of △ABC. If AD = AE and BD = CE, prove that AB = AC.* [MAYHEM problem HS33]

Solution. Since $AD = AE$, $DB = EC$ and $\angle BAC$ is common to triangles ABD and ACE, and we also know that D lies between A and C and E between A and B, we must have $\triangle ABD$ congruent to $\triangle ACE$, so $AB = AC$ as required. (The fact that D lies on AC and E on AB is important since otherwise there would be two possible orientations of $\triangle ABD$ given $AD = AE$, $DB = EC$ and $\angle BAC$.)

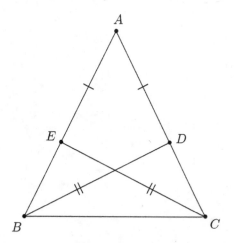

Problem 11.182. *Two circles C_1 and C_2, with centres O_1 and O_2, are externally tangent. PQ is a common tangent to the circles, with P on C_1 and Q on C_2, and M is the mid-point of PQ. Show that $\angle O_1MO_2 = 90°$.* [MAYHEM problem HS64]

Solution. Let R be the common point of tangency of the two circles. Draw the tangent at R, and let it intersect PQ at N. By the circle tangent theorem, $NP = NQ = NR$. Since M has been defined as the mid-point of PQ, $N = M$. Since MP and MR are tangents to C_1, it follows that O_1M bisects $\angle PMR$. Similarly, O_2M bisects $\angle RMQ$. Therefore, $\angle O_1MO_2$ is half of $\angle PMQ$, or 90°.

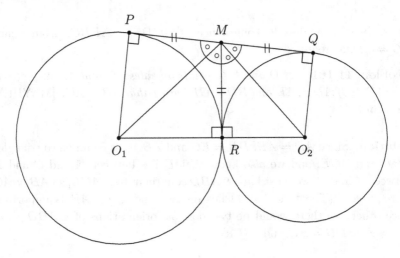

Index